Expert Systems: Applications
to Urban Planning

T.J. Kim   L.L. Wiggins   J.R. Wright
Editors

Expert Systems: Applications to Urban Planning

With 48 Illustrations

Springer-Verlag
New York Berlin Heidelberg
London Paris Tokyo Hong Kong

T.J. Kim, Department of Urban and Regional Planning, University of Illinois, Champaign, Illinois 61820, USA

L.L. Wiggins, Department of Urban Studies and Planning, Massachusetts Institute of Technology, Cambridge, Massachusetts 02139, USA

J.R. Wright, Department of Civil Engineering, Purdue University, Lafayette, Indiana 47907, USA

Library of Congress Cataloging-in-Publication Data
Kim, Tschangho John.
    Expert systems : applications to urban planning / edited by
Tschangho John Kim, Lyna L. Wiggins, and Jeff R. Wright.
        p.   cm.
    ISBN 0-387-97171-8
    1. City planning—Data processing.   2. Expert systems (Computer
science) I. Wiggins, Lyna L.   II. Wright, Jeff R.   III. Title.
HT166.K5545   1990
711′.4′0285633—dc20   89-37100

Printed on acid-free paper.

Camera ready copy prepared by the editors.
Printed and bound by Edwards Brothers, Inc., Ann Arbor, Michigan.
Printed in the United States of America.

9 8 7 6 5 4 3 2 1

ISBN 0-387-97171-8 Springer-Verlag New York Berlin Heidelberg
ISBN 3-540-97171-8 Springer-Verlag Berlin Heidelberg New York

Preface

Urban Planning is a multi-dimensional and multi-disciplinary activity embracing social, economic, political, and technical factors. Solutions for urban problems frequently require not only numerical analysis but also heuristic analysis, which in most cases depends on the planners' intuitive judgement. Conventional quantitative techniques, that deal mainly with numerical analysis, lack the capability of incorporating the heuristic knowledge of planners into problem solving.

While expert systems have become a popular topic in the computing, medical, and engineering fields, the expert system is still a new technology in urban planning. This book will introduce expert systems for problem solving in urban planning and will describe the way in which heuristic knowledge and rules of thumb of expert planners can be represented through computer programs. This book will present practical applications of expert systems for solving many important urban planning problems, particularly those issues that many practicing planners face in their daily operations.

The book is organized into five distinct, but complementary parts: (I) An Introduction to Expert Systems, (II) Applications to Land Use and Transportation Planning, (III) Expert Systems in Site Selection, (IV) Expert Systems in Environmental Planning, Conflict Mediation and Legal Disputes, and (V) Future Research Directions. Part One will introduce the applicability of expert systems to urban planning problem solving. This material will be presented in two chapters—one summarizing expert systems in urban planning, the second presenting a thorough overview of the technology and tools for systems development in detail.

Parts Two, Three, and Four contain collections of papers organized by domain of application and grouped by complexity of implementation. In this way, the reader may elect to read the initial papers of each section to acquire insight about more elementary and proven applications, to select a domain of particular interest and acquaint him/herself with a breadth of implementations ranging from straightforward production systems to complex systems developed using high-end development tools, or to review a range of domain applications produced with a specific tool or class of tools, possibly for purposes of comparison with ones with which the reader may be familiar. This organization is intended to meet the needs of a wide variety of readers from the most experienced expert systems researcher seeking ideas for new systems design, to domain experts wishing to gain insights as to how this important technology might be applied to their field of expertise. The final part of the book presents future developments and directions.

This book is the collection of recently completed works by active researchers in expert systems applications to urban planning. It is our hope that it will prove to be of value to a number of different kinds of readers: urban planners, regional scientists, geographers, economists, systems analysts, computer scientists, operations researchers, and social scientists who are concerned with computer applications to solving problems of cities and regions. Permission to quote and to reproduce figures, tables, and materials has been kindly granted by the American Planning Association, Pion Limited, the American Society of Civil Engineers, Pergamon Press, Inc., and Elsevier Publishing Co. We also wish to thank the editors of the several journals involved who graciously permitted us to reprint all or portions of the included articles.

We are indebted to Ms. Kamie Redinbo for preparation of this manuscript. Her skill is impressive; her patience, inspiring.

Tschangho John Kim
Lyna L. Wiggins
Jeff R. Wright

Contents

PART THREE

PART FOUR

PART FIVE

The Editors

Tschangho John Kim is a Professor in the Department of Urban and Regional Planning at the University of Illinois at Urbana-Champaign. For twelve years, he has been teaching computer applications to urban and regional planning, transportation planning, land use planning, and economic analysis of public plans and policies. Dr. Kim has also been actively involved in research in solving large scale planning problems using supercomputers. Current research projects include building expert systems for the coal trade and facility siting, and developing an efficient solution algorithm for non-linear programming models. Dr. Kim holds a bachelor's degree in Architectural Engineering and both a MCP and a Ph.D. degree in Urban and Regional Planning from Princeton University. His publications include a book, *Integrated Urban Systems Modeling: Theory and Applications*, and numerous articles and reports in urban, regional and national planning.

Lyna L. Wiggins is Assistant Professor of Urban Studies and Planning at the Massachusetts Institute of Technology where she teaches courses in quantitative methods, computer applications in planning and utility systems planning. Her research has included development of demand forecasting models and facility siting methods for energy facilities. She has designed decision support systems for environmental planning and infrastructure planning and management. Recent research has included an examination of the adoption of microcomputer technology in urban and regional planning agencies, and an analysis of organizational factors on system implementation and effectiveness. Dr. Wiggins holds bachelor's degrees in mathematics and city and regional planning from California Polytechnic State University, San Luis Obispo, a master's degree in statistics from Stanford University and a Ph.D. in urban and regional planning from University of California, Berkeley.

Jeff R. Wright is Associate Professor of Civil Engineering at Purdue University where he teaches and conducts an active research program focused on the application of computer technology to the solution of engineering management problems. Dr. Wright has designed and implemented decision support systems for a wide variety of public-sector engineering problems including facilities location, scheduling and sequencing, route design and management, spatial design and analysis, and systems operations. His present research interests include the integration of knowledge-based procedures into geographical information systems. Dr. Wright holds bachelor's degrees and a Masters in Civil Engineering, Social Psychology and Civil Engineering, respectively from the University of Washington, and a Doctorate from The Johns Hopkins University Whiting School of Engineering in Systems Analysis and Economics for Public Decision Making.

Contributors

Kimberley Adams is a systems engineer at Corning Glass Works in Corning, New York. She is involved in developing all phases of expert systems developments at Corning Glass Works. She received her MS in General Engineering and Operations Research from the University of Illinois at Urbana-Champaign.

Dr. Richard K. Brail is an Associate Professor, and the Chair and Graduate Director of the Department of Urban Planning and Policy Development at Rutgers University. He is also the director of the Regional Policy Simulation Laboratory at Rutgers. He is the author of a book, Microcomputers in Urban Planning and Management, and numerous articles in the areas of urban transportation and computer applications in planning. He has been active in both research and consulting in the areas of computer applications and transportation modeling for over 20 years. Dr. Brail received a B.A. in philosophy from Rutgers University, and both MCRP and Ph.D. degrees in city and regional planning from the University of North Carolina, Chapel Hill.

Paul Der-Ming Cheng is currently a Research Assistant Professor at Vanderbilt University. He holds an M.S. in Transportation from the National Taiwan University and a Ph.D in Civil Engineering from Rensselear Polytechnic Institute. His current research interests include the routing of hazardous materials and the development of highway planning and engineering models.

Shih-Miao Chin is a research staff member at the Oak Ridge National Laboratory (ORNL). He holds a B.S. in mathematics from the National Taiwan University, a Masters degree in mathematics from Utah State University, a Masters degree in Civil Engineering from the University of Utah and a Ph.D in Civil Engineering from Rensselear Polytechnic Institute. He has worked as a transportation engineer with the Utah Department of Transportation and with Greenman-Pedersen Inc. His areas of interest include traffic flow simulation, computerized traffic signal control, interactive computer graphics, and expert system software applications.

James Richard Davis is a Principal Research Scientist with the CSIRO Division of Water Resources in Canberra, Australia. He leads a group concerned with applying knowledge-based systems to natural resource planning and management problems. The group has developed expert system shells and applied them to such problems as predicting the effects of fires in a major Australian national park, estimating the risk from tick-born diseases in cattle and predicting the trafficability of remote parts of Australia under varying seasonal conditions. He has also developed knowledge-based programs for land use planners and applied them to practical planning problems. Current projects include the development of a combined expert system and GIS for the land management of a major Australian Army base and a spatial decision support system for analyzing the effects of land management policies on water quality in the catchment of a major Australian city. His recent publications include a bibliography of expert systems in natural resource management and chapters in books on computer tools for planners. Dr. Davis holds a B.Sc (Hons) in physics from Otago University, NZ, and a bachelors degree in economics and a PhD in nuclear physics from the Australian National University.

Irene T. Findikaki is a civil engineer now working with Bechtel Corporation as a computer engineering application specialist. During the past 5 years she has worked on geographic information systems and groundwater modeling applications for remediation of hazardous waste sites. Her research interests include work in expert systems, numerical modeling, geographic information systems, and computer graphics. Dr. Findikaki holds a bachelors degree in engineering/architecture from the National Technical University of Athens, a M.S. in urban and regional planning from San Jose State University, and a Ph.D. in civil engineering from Stanford University.

Ian W. Grant is an Experimental Scientist now with the CSIRO Division of Water Resources in Perth, Australia. He was previously with the Division's Knowledge Systems group in Canberra, where, apart from writing the ADAPT knowledge based planning program, he wrote a spatial inferencing expert system and routines for geographic data processing. Currently he is using SMALLTALK, to write a program to classify and analyze arguments used by interest groups concerned about resource distribution. His research interests include computational geometry, rhetorical structure and application of artificial intelligence to problems of natural resource management. Mr. Grant holds a B.Sc.(Hons) in Computer Science from Macquarie University, Sydney.

Sang-Yun Han is a Ph.D candidate in the Department of Urban and Regional Planning, University of Illinois at Urbana-Champaign. Mr. Han holds a Bachelor's degree in Political Science from University of Iowa, and a Master of Urban Planning degree from University of Illinois at Urbana-Champaign. His Ph.D dissertation is on expert urban information systems.

Eric J. Heikkila is assistant professor of urban and regional planning at the University of Southern California where he teaches courses on quantitative methods and policy analysis. His primary research focus is on the interplay between urban economic, geographic and demographic structures, and he has authored many scholarly articles in these areas. Prior to his current position, Dr. Heikkila worked for five years as an urban planner and financial analyst for the City of Vancouver. He received his doctorate in economics from the University of British Columbia.

Roozbeh Kangari is Associate Professor of Civil Engineering at Georgia Institute of Technology where he teaches courses in expert systems, robotics, and database management in civil engineering. Dr. Kangari has done extensive research in the area of risk management and knowledge-based expert systems in construction management. His currently completed research projects, funded by the National Science Foundation, include application and feasibility of robotics in the construction industry, and fuzzy sets and expert systems in risk analysis. He is involved in research with the Computer Integrated Manufacturing Systems (CIMS) at Georgia Tech. Dr. Kangari holds a Ph.D. degree in Civil Engineering (Specialized in Construction Engineering and Management) from University of Illinois at Urbana-Champaign. He has published numerous technical papers, and is a member of ASCE, IEEE, and PMI.

Moonja Park Kim is a Principal Investigator at the Construction Engineering Research Laboratory, U.S. Army Corps of Engineers in Champaign, Illinois. Dr. Kim has extensively worked as a Knowledge Engineer in developing expert systems in the construction management field. She holds a Bachelor's degree in Psychology from Douglass College and a Master's and Ph.D in Psychology from Rutgers University. She is a member of AAAI, SAME, and a technical committee member of Microcomputers in Construction, Construction Division of ASCE.

Yi-Chin Lee is a civil engineer now with Oracle Corporation in Belmont, California, where he is designing and implementing management information systems for the manufacturing and financial services industries. His dissertation research, in the area of expert systems in planning, was the design and implementation of a machine-based meeting environment that utilizes artificial intelligence techniques, decision analysis, and statistical analysis to help resolve environmental disputes in facility siting. His other research interests include energy resource planning, environmental dispute resolution, and data base management. Dr. Lee holds a B.S. in civil engineering from National Taiwan University, and a Masters and Ph.D. in civil engineering from Stanford University.

James E. Moore is Assistant Professor of Urban and Regional Planning and Civil Engineering at the University of Southern California, Los Angeles, where he teaches courses in Planning, Civil Engineering, and Industrial Engineering. For the past six years, Dr. Moore has been doing fundamental research on the engineering and economic aspects of large-scale transportation and land use systems. His current research involves the mathematical definition of urban subcenter, the efficiency of Transferrable Development Rights markets, the use of expert systems as training sets for neural networks, and the use of neural networks to predict equilibrium traffic flows in transportation networks. Dr. Moore holds a Bachelor's degree in Industrial Engineering and Urban and Regional Planning, and a Master's degree in Urban and Regional Planning from Northwestern University; a Master's degree in Industrial Engineering from Stanford University; and a Doctorate in Civil Engineering, also from Stanford.

Leonard Ortolano, UPS Foundation Professor of Civil Engineering at Stanford University, specializes in water resources and environmental planning. His current research includes work on environmental management in developing countries and expert systems in environmental engineering. He has worked on expert systems for maintaining sewer networks and for operating both water and wastewater treatment plants. Dr. Ortolano holds a Bachelor's degree in Civil Engineering from the Polytechnic Institute of Brooklyn and a Master's and

Ph.D. in Engineering from Harvard University. His publications include the textbook, *Environmental Planning and Decision Making*, published by John Wiley & Sons in 1984.

Catherine D. Perman is a civil engineer now consulting for major automotive companies and government agencies in the development of model-based reasoning systems. Her dissertation research, in the area of expert systems in environmental engineering, was the design and implementation of a prototype expert system to assist wastewater treatment plant operators diagnose operating problems and recommend remedial action. Dr. Perman also has 5 years of experience in environmental engineering, regulatory analysis, and mathematical modeling for a major oil company. She holds a B.A. in geology/applied math from Brown University, a M.S. in civil and environmental engineering from Utah State University, and a Ph.D. in civil engineering from Stanford University.

Shahrokh Rouhani is Assistant Professor of Civil Engineering at Georgia Institute of Technology where he teaches courses in hydrology, groundwater hydrology, and water design, space-time mapping of geohydrological variables, and application of expert systems in water resources planning and management. Recent research has included the design of regional groundwater quality monitoring networks and multivariate analysis of spatiotemporal data. He has published numerous technical papers, and is a member of ASCE, AGU, and IWRA. Dr. Rouhani holds Bachelor's degrees in Civil Engineering and Economics from University of California, Berkeley and a Master's degree and a Ph.D. in Engineering Sciences from Harvard University.

Frank Southworth is Leader of the Transportation Operations Research and Planning Group at the Oak Ridge National Laboratory (ORNL). He holds a B.A. with Honors and a Ph.D in Geography from the University of Leeds, England. Prior to joining ORNL in January 1984, he was a research officer in the Institute for Transportation Studies, University of Leeds and was on the faculty of the Civil Engineering Department at the University of Illinois. His current research includes algorithm design for global airlift scheduling, real time highway traffic monitoring, national highway network design, crisis mobility planning, and an assessment of the transportation sector's contributions to the "greenhouse effect".

Sunduck Suh is a Senior Researcher at the Korea Transport Institute. Dr. Suh holds a Bachelor's degree in Civil Engineering and Master's degree in Civil Engineering, both from Seoul National University and received his Ph.D in Regional Planning from the University of Illinois at Urbana-Champaign. He is currently working on developing expert transport planning systems for Korea.

PART ONE

Introduction to Expert Systems

Introductory remarks by Lyna L. Wiggins

The artificial intelligence (AI) community has been associated with research into robotics, natural language processing, and the ability of computers to manipulate non-numerical symbols. Expert systems are the most widespread commercial applications to come from the AI research laboratories. Expert systems that are operational outside of the research facilities are quite recent, and it is exciting to be working in a field where the "Dark Ages" are less than a decade in the past. However, since the field of expert systems is growing and changing rapidly, there is also some difficulty in keeping up with the latest terminology and technology. The chapters in Part One are intended to help the reader with little knowledge of the expert systems literature gain the background to read the later application chapters with more fluency.

In Chapter 1, Ortolano and Perman provide an overview of expert systems and their application to urban planning. They begin with definitions and examples, and emphasize the features of expert systems which make them different from traditional programs. Next they describe a general classification of expert systems by the types of problem-solving activities they perform. This classification includes systems for interpretation, diagnosis and prescription, design and planning, monitoring and control, and instruction. For each class of problem some existing planning applications of expert systems are described. Most of the applications described in the later chapters of this book will be systems created to address design and planning problems. The authors emphasize that, to be appropriate for an expert system, a problem must be well-selected. They outline six possible criteria for deciding whether a particular problem will be amenable to an expert system solution. Since many planning and design problems will not fulfill these criteria, this is an important section.

In Chapter 2, Wiggins, Han, Perman and Lee provide an introduction to expert systems technology. The authors introduce concepts and terms in common-use in the field, and are careful in this chapter to provide definitions of "jargon" for the reader. The chapter begins with a discussion of the software tools available for expert systems development, including a variety of languages and shells. Most of the applications in the later chapters of this book have been developed using shells developed by commercial software vendors. The five shells described in Chapter 2 are representative of the range of software currently available to expert systems developers. For each shell there is a discussion of the required hardware configuration, the editing function, the control mechanism, the line of reasoning choices available, and its advantages and disadvantages. The chapter ends with a discussion of specific criteria that planners may find of concern in their selection of software.

CHAPTER 1

Applications to Urban Planning: An Overview[1]

Leonard Ortolano and Catherine D. Perman

Since the early 1980s expert systems have received much attention from the news media, from programmers, and from managers and decision makers.[1] Planners' interest in the potential of computers to aid planning practice has sparked their curiosity about expert systems. Although expert systems to aid planners are not yet operational, research is underway into applications such as site analysis and zoning ordinance implementation. This article introduces expert systems, surveys their applications for planning, and explores how useful they may become in urban and regional planning and related fields.

Expert systems represent one component of a larger effort in programming computers to perform tasks that require intelligence. Early work on this evolved into the branch of computer science now known as "artificial intelligence." Besides expert systems, the artificial intelligence field includes applications such as robotics and language interpretation.

What Are Expert Systems?

Expert systems are computer programs that apply artificial intelligence to narrow and clearly defined problems. They are named for their essential characteristic: they provide advice in problem solving based on the knowledge of experts. Expert systems typically combine rules with facts to draw conclusions; the process relies heavily on theories of logical deduction developed by

---

1   The original version of this chapter was published in *The Journal of the American Planning Association*, 1987, 53(1):98-103.

mathematicians and philosophers, and adapted to particular applications by engineers, scientists, planners, and managers across a wide range of disciplines.

Both heuristic methods and conventional computer programs (e.g., FOR-TRAN programs) are often used in expert systems. Heuristic methods include rules of thumb, intuitions, simplifications, judgments, and other problem-solving approaches that may not find the best solution but often find useful solutions quickly.

The subject area of an expert system, such as site planning or zoning administration, is called its domain. The collection of facts, definitions, rules of thumb, and computational procedures that apply to the domain is called its knowledge base. Sources of that knowledge include published Materials, quantitative analysis programs, and the intuitions and problem-solving strategies of experts in the subject area.

The set of procedures for manipulating the information in the knowledge base to reach conclusions is called the control mechanism. A control mechanism might be based, for example, on logical deduction of conclusions from a set of facts and rules of the form "If (premises), then (consequences)." Axioms and theorems of symbolic logic are used to deduce conclusions from rules and facts about the premises of various rules. The control mechanism includes procedures for determining which rules to examine first and which facts to obtain by querying the user.

Besides the knowledge base, control mechanism, and user interface, an expert system typically has a working memory that contains information generated during a particular run of the system. The system also may include facts and other information obtained from conventional numerical analysis programs, data bases, or remote sensors. (Generally, the latter sources need special interfaces for use with the system.)

The following example clarifies the distinctions between the knowledge base, control mechanism, and working memory. Consider a knowledge base that includes these rules:

```
If the proposed building is residential
and the building will be in an area zoned R1,
then subsection 10.1 of zoning ordinance is applicable.

If the proposed building is residential,
then section 10 of the zoning ordinance is applicable.
```

Suppose the working memory includes the fact that the proposed building is residential and that it will be located in an area zoned R1. Each of the above rules in the knowledge base is applicable, and the expert system must be instructed how to determine which rule to invoke. The decision can be made using a control mechanism that includes the following general strategy: if all the premises of two or more rules are satisfied, invoke the rule with the greatest number of premises. This procedure uses more of the data in the working memory and yields a more specific consequence.

A key feature of expert systems is that the knowledge base is coded separately from the control mechanism. This separation, which distinguishes expert systems from conventional programs, is important for several reasons.[2] First, it is the basis for generic software known as expert systems shells (or frameworks). A shell consists of a general control mechanism and an editing facility for entering the knowledge base for a particular domain. Numerous shells are available at a variety of prices and levels of sophistication.[3]

Second, the independence of the knowledge base from the control mechanism enables the user to add to the knowledge base without rewriting substantial portions of the computer program. Often this involves simply coding additional "if-then" rules into the knowledge base. This ease of expanding or modifying the knowledge base makes it possible to use expert systems to solve problems through rapid prototyping. In this process the user codes the initial prototype solution based on very incomplete knowledge. Experts' reactions to runs of the initial prototype program yield information that enlarges and improves the knowledge base. In an iterative process, runs of successive prototypes clarify how the knowledge base can be improved further and the next prototype thus refined. The user can perform each iteration without rewriting substantial portions of the computer program.

Another feature that distinguishes expert systems from conventional computer programs is their transparency: they allow the user to ask questions during an interactive session to learn how the program reached a conclusion or why the program asked the user for particular information. A user who asks why a certain conclusion is true will be shown the chain of reasoning the program used to reach the conclusion. If the program asks for a particular fact and the user asks why the fact is needed, the program will explain the line of reasoning it is pursuing and why the information requested is important.

The programming language used to implement an expert system is not its defining characteristic. Although most expert systems research is done in LISP or PROLOG, languages that process character strings efficiently, it is possible to develop expert systems using BASIC or FORTRAN.

Expert Systems Applications

Until recently there were applications for expert systems in only a few fields. Before 1980, most of the expert systems were in medicine and chemistry. Since then there has been a rush to build expert systems in a wide range of areas.[4] Many of the existing systems are in the research stage, and relatively few have been refined and tested to where they are used routinely in field settings. That is changing rapidly, however, as more and more private companies investigate applications that have commercial potential.[5]

Applications of expert systems can be described according to which of the following generic problem-solving activities they perform: interpretation, diagnosis and prescription, design and planning, monitoring and control, and instruction.[6] A particular expert system may involve several types of problem-solving activities; this categorization, while commonly used, is neither unique nor complete. Below we discuss potential applications of each activity for urban and regional planning. We provide examples of each and/or give information on current research into and availability of expert systems in each area.

Interpretation means inferring situation descriptions from data. There have been numerous applications involving this activity, and several have promise for planning:

- *Combine expert systems with traditional data base management programs to create an "intelligent" data base that can aid in retrieving data for subsequent analysis.* An intelligent data base includes a "front end" that incorporates the knowledge of an expert in the domain of the data base in order to help users formulate queries, devise strategies for efficient searches of the data base, and eliminate inconsistencies and repetitions in retrieved data.[7] Although no urban planning examples of intelligent data bases are operational yet, this is an active area of research. For example, Tanic's (1986) prototype "intelligent urban information system" for Fort-de-France in Martinque combines an expert system with a traditional data base management program. Another example is in research at Carnegie-Mellon University to develop an intelligent data base for the Electric Power Research Institute in Palo Alto, California.[8] Other work on intelligent data base systems is reported by Kerschberg (1986).[9]

- *Assist in analyzing land use laws and other legal issues.* Waterman's summary (1985, 267-270) of 10 expert systems in this area evidences the potential applicability of expert systems to legal issues of importance to planners. Topics covered by those systems include issues faced by contractors, auditors, and tax planners.

- *Determine whether a proposed land use meets zoning and other local and land use regulations.* Expert systems that aid in the implementation of laws and regulations already have received considerable attention in fields other than urban planning. According to Harmon and King (1985, 100), a firm called Expert Systems International[10] has developed expert systems that provide advice on building regulations. Researchers at Oxford Polytechnic are in the early stages of a project to build an expert system to aid local planners in implementing development controls (personal communication with Michael Leary, Department of Town Planning, Oxford Polytechnic, United Kingdom, May 7, 1986).

- *Estimate probable damage to property in the event of natural catastrophes.* Waterman (1985, 261) has abstracted the extensive work on building safety and damageability conducted at Purdue University.

Diagnosis and prescription means inferring malfunctions from observable data and prescribing remedies. Although many expert systems have been developed in medicine and engineering to perform this function, relatively few such applications are envisioned for urban planning. Nevertheless, numerous expert systems in this category will be of interest to local governments concerned with public works:

- *Identify causes of problems in managing particular projects.* An example of such a project management expert system is summarized in Waterman (1985, 266). Several expert systems have been developed for the management of construction activities (see, for example, Levitt and Kunz, 1985).

- *Assist in operation and maintenance of physical infrastructure systems such as roads, sewers, and wastewater treatment facilities.* For examples involving operations of electric power facilities, see Electric Power Research Institute (1985a and 1985b). Expert systems to aid in highway, bridge, and sewer system maintenance have been developed by Ritchie and Mahoney (1986), Maser (1986), and Destrigneville et al. (1986), respectively.

Design and planning applications are those that design the form or arrangement of objects and actions under given constraints. Some expert systems research has been initiated in site analysis and land use location, and more can be expected:

- *Perform analysis of a particular site.* A specialized example of an expert system for site analysis—an area of long-standing interest to urban and regional planners—is the effort by Law, Zimmie, and Chapman (1986) to characterize inactive hazardous waste disposal areas. Of related interest is the work combining expert systems with computer-aided design and drafting packages (see, e.g., Gero and Coyne, 1986).

- *Find locations well suited for a specific land use.* An initial effort to develop an expert system to site a particular land use is presented by Findikaki (1986).

- *Assist in planning controlled fires to manage habitats in large parks.* An expert system of special interest to park planners and environmental specialists is the one developed to assist in managing a large national park in Australia. The system is notable because it has been field-tested, embodies knowledge of ecology, and uses zones to break down a vast area for analysis. An extensive body of work on expert systems in park management is reported by Williams, Nanninga, and Davis (1986).

- *Support mathematical modeling by providing assistance in selecting, calibrating, and interpreting the output of complex models.* Expert systems of this type are notable since mathematical models are commonly used in urban and regional planning. An often-cited example of an expert system for deciding which of several complex mathematical models should be used in a given situation is SACON, developed by Bennett et al. (1978) to assist in selecting models for structural analysis. Calibration is illustrated by HYDRO, an expert system that assists in calibrating a very complex watershed hydrology model (see Gaschnig, Reboh, and Reiter, 1981). Output interpretation is demonstrated by the AT&T Bell Laboratories system that interprets numerical outputs for users of a standard statistical analysis computer program (see Rauch-Hinden, 1985, 141-42).

Monitoring and control means comparing observations to expected outcomes and governing the behavior of the monitored system. Applications of this type are likely to be of great interest to public works specialists but of only marginal interest to urban and regional planners:

- *Incorporate expert systems on microchips in instruments to control environmental conditions within a building.* EPRI has received an unsolicited proposal for a "smart building" that would be capable of reacting to time-of-day changes in electricity prices and anticipated weather conditions (personal communications with Shishir Mukherjee, project manager, EPRI, Palo Alto, California, November 14, 1985). Local governmental units that operate public works facilities would find value in expert systems developed for real-time system control (see e.g., Rauch-Hinden 1985, 125-29, on power plant facility controls and Zozaya-Gorostiza and Hendrickson, 1986, on traffic signal controls).

Instruction. The final activity category, instruction, defines expert systems that help novices understand concepts or learn to perform specific tasks:

- *Teach new personnel about complex zoning requirements, development control laws, and the like.* It has been argued that, once an expert system is built for a particular task, it may be possible to modify the system to create an instrument for training novices in how to perform that task.[11] Thus, many of the expert systems applications described above may be useful in the design of systems for instructional purposes.

Selecting the Right Task for an Expert System

Building an expert system to aid in performing a particular task is neither quick nor inexpensive. One way to avoid wasting resources in developing an expert system is to make sure, at the outset, that the task is one an expert system can do productively. Examples of the meaning of the word "task" are: deciding whether to issue a building permit for a particular project; finding a suitable location for a proposed industrial facility; and estimating the expected benefits of improving a building's ability to resist earthquakes. People experienced in developing expert systems often cite six conditions for deciding whether the knowledge used to perform a particular task may be effectively codified in an expert system:[12]

1. The knowledge needed for task performance is specialized and narrowly focused, such as the knowledge of a city attorney checking on a development project's compliance with local ordinances. Many of the attorney's analysis procedures can be stated precisely enough to be programmed.

2. True "experts" exist, in the sense that individuals identified as experts perform the task better than novices. If that is not the case, there is little to be gained by building an expert system, since it will not improve the performance of the nonexpert.

3. The task is neither trivial nor exceedingly difficult. The appropriate range of difficulty is usually defined as a task that takes an expert more than 15 minutes but less than several hours to perform. For a planner just learning to build an expert system it makes sense to select a relatively easy task. If the task is too simple, however, the expert system will be superfluous.

4. Conventional computer programs (using FORTRAN, spreadsheets, etc.) are inadequate for the task. If the user can get the help he or she needs with conventional (algorithmic) programs, he or she should use those programs. If task performance involves rules of thumb, symbolic reasoning, or other non-algorithmic approaches, an expert system might be helpful.

5. The potential payoff from an expert system is significant. High payoffs can involve efficiency gains such as substantially improved task performance or decreased costs. Payoffs that have been reported include making limited expertise more widely available and training novices in the procedures followed by experts. Common motives for building an expert system include the need to codify the expertise of a specialist who is about to leave an organization or whose services are needed frequently and simultaneously at many widely separated locations. An added gain of building an expert system is an enhanced understanding of the task itself. An expert's understanding often improves with the experience of codifying his or her knowledge.

6. An articulate expert is available and willing to make a long-term commitment to helping build the expert system. The expert must be prepared to describe his or her reasoning in performing the task and to critique the prototype expert systems developed during the system building process. The time required to build an expert system varies with the complexity of the task, but it is not uncommon for system development to require at least two years (Waterman, 1985).

Although it is true that many urban planning problems are poorly defined and institutionally complex, that doesn't mean expert systems will have no role to play in planning. Instead it underscores how important it is that planners restrict their efforts in building expert systems to tasks that satisfy the conditions elaborated above.

Will There Be Expert Systems to Aid Planners?

There is no question that attempts will be made to develop expert systems for urban and regional planners. At least some researchers believe expert systems are suitable, based on the above criteria, for a number of tasks of interest to planners. Prototype systems of potential value to planners are being built now.[13]

Harder questions concern whether the expert systems for planners consistently will yield the same level of performance as experts do and whether they will be acceptable to the planners they are designed to assist. These difficult questions are debated regularly at conferences on expert systems and are associated with the criticism that expert systems are "nothing new" and little more than "media hype." It is impossible to determine now whether expert systems will live up to the claims of the many firms building them. As Hewett (1986) suggests, the investment of hundreds of millions of dollars in expert systems by companies and governments is based largely on the promise of prototypes, not on the evidence provided by fully tested, operational systems.[14] The two-year interval typically required to develop a substantive, user-friendly prototype accounts partly for the paucity of operational systems. Moreover, questions of system validity and legal liability for system "errors" have not received extensive attention except from a few researchers.[15]

If the development of expert systems applications in urban and regional planning follows the pattern established earlier in medicine, chemistry, and engineering, the next few years will witness the introduction of many promising but untested prototype systems for planners. It will be longer before the validity and utility of those prototypes for planning can be thoroughly assessed. Only in time will we know whether expert systems will be valuable to planners. For now, the promise of these systems is substantial enough to have attracted the attention of numerous researchers, governments, entrepreneurs, and potential clients. Planners should keep abreast of developments in this fast-paced field.

Notes

1. A comprehensive and readable introduction to expert systems is given by Harmon and King (1985).
2. This discussion of differences between expert systems and conventional computer programs is based on Fenves (1986).

3. Guilfoyle (1986) and Harmon and King (1985) provide introductions to expert system building tools.

4. Waterman's survey (1985) includes abstracts of more than 180 expert systems, roughly 85 percent of which were described in materials published during the 1980s. The seven most active areas of expert systems research were chemistry, computer systems, electronics, engineering, geology, medicine, and the military.

5. Based on a survey of more than 60 companies engaged in expert systems work in North America, Hewett (1986) reports: "At least 1,000 major projects are under way covering a wide range of applications in a variety of industries. . . While few systems are yet operational, the leading developers of expert systems have proved to themselves, as a result of developing successful prototypes, that the technology will produce worthwhile payoffs in the future."

6. A review of the examples surveyed by Waterman (1985) indicates that there have been numerous applications involving interpretation, diagnosis and prescription, and monitoring and control. A smaller but nontheless notable number of applications involve planning and design and instruction.

7. An intelligent data base having these features is described by Jakobsen et al. (1986).

8. Shishir Mukherjee of EPRI is project manager for this study, "An Intelligent Information Center."

9. The process of creating an intelligent data base is facilitated by expert systems shells that are elements of integrated microcomputer packages that include data base managers and word processors. An illustration of such a software product is GURU, available from Micro Data Base Systems, Inc., P.O. Box 248, Lafayette, IN 47902. GURU is a a trademark of Micro Data Base Systems, Inc.

10. The firm's U.S. office is located at 1150 First Avenue, King of Prussia, PA 19406.

11. Harmon and King (1985, 242-44) make this point in the context of a medical diagnosis training program called GUIDON. The program was built using much of the knowledge base developed for MYCIN, a widely known expert system to assist in diagnosing infectious blood diseases.

12. The six conditions were derived from Fenves (1984), Prerau (1985), Waterman (1985), and Buchanan et al. (1983).

13. Examples of prototype systems in advanced stages of development include the construction project management system by Levitt and Kunz (1985) and the park management system by Williams, Manninga, and Davis (1986). The latter actually has been used by park managers in Australia.

14. Hewett's statistics (1986) on expert systems activity in North America are informative. He estimated that commercial users would invest $300 million in "expert system technology" in 1986. Based on his interviews with 14 leading users of expert systems technology (with a combined staff of 500 people working on more than 100 expert systems products), not one company had a "fully operational [expert systems] application in late 1985."

15. For an extensive analysis of the validity of particular expert systems, see Buchanan and Shortliffe (1984). A discussion of legal liability issues related to the use of expert systems is given by Bundy and Clutterbuck (1985).

References

Bennett, J., L. Creary, R. Englemore, and R. Melosh, 1978. "A Knowledge-Based Consultant for Structural Analysis", Stanford, Calif.: Department of Computer Science, Stanford University.

Buchanan, B.G., D. Barstow, R. Bechtel, J. Bennett, W. Clancy, et al., 1983. "Constructing an Expert System", in *Building Expert Systems,* edited by F.Hayes-Roth, D.A. Waterman, and D.B. Lenat. Reading, Mass.: Addison-Wesley.

Buchanan, B.G., and E.G. Shortliffe, eds., 1984. *Rule-Based Expert Systems,* Reading, Mass.: Addison-Wesley.

Bundy, A., and R. Clutterbuck, 1985. "Raising the Standards of AI Products", in *Proceedings of the Ninth International Joint Conference on Artificial Intelligence,* Vol. 2. Proceedings of a conference held in Los Angeles, August 18-23.

Destrigneville, B., P. LeCloitre, G. LeCoeur, D. Trayaud, and R.S. Macgilchrist, 1986. "Sewerage Rehabilitation Planning Expert System", in *Eau et informatique, actes du colloque organise par l'Ecol Nationale des Ponts et Chaussees,* Proceedings of a conference held in Paris, France, May 28-30.

Electric Power Research Institute, 1985a. "BWR Shutdown Analyzer Using Artificial Intelligence Techniques", Special report. Palo Alto, Calif.: EPRI.

Electric Power Research Institute, 1985b. "Functional Specifications for AI Software Tools for Electric Power Applications. Report under project RPZ582-01. Palo Alto, Calif.: EPRI.

Fenves, S.F., 1984. "Artificial Intelligence-Based Methods for Infrastructure Evaluation and Repair", in *Infrastructure: Maintenance and Repair of Public Works,* edited by A.H. Nolof and C.J. Turkstra. Annals of the New York Academy of Sciences, Vol. 431 (December 5). New York: New York Academy of Sciences.

Fenves, 1986. "What is an Expert System?" in *Expert Systems in Civil Engineering,* edited by C.L. Kostem and M.L. Maher, New York: American Society of Civil Engineers.

Findikaki, I.T., 1986. "An Expert System for Site Selection", in *Expert Systems in Civil Engineering,* edited by C.L. Kostem and M.L. Maher. New York: American Society of Civil Engineers.

Gaschnig, J., R. Reboh, and J. Reiter, 1981. "Development of a Knowledge-Based Expert System for Water Resources Problems", *Report SRI 1619.* Menlo Park, Calif.: AI Center, SRI International.

Gero, J.S., and R.D. Coyne, 1986. "Developments in Expert Systems for Design Synthesis", in *Expert Systems in Civil Engineering,* edited by C.L. Kostem and M.L. Maher. New York: American Society of Civil Engineers.

Guilfoyle, C., 1986. "A Table Load Full of Micro Shells", *Expert Systems User,* April: 18-20.

Harmon, P., and D. King, 1985. *Expert Systems: Artificial Intelligence in Business,* New York: John Wiley and Sons.

Hewett, J., 1986. "Commercial Expert Systems in North America," in *Proceedings of the Sixth International Workshop on Expert Systems and Their Applications,* proceedings of a conference held in Avignon, France, April 28-30.

Jakobsen, G., C. Lafond, E. Nyberg, and G. Piatetsky-Shapiro, 1986. "An Intelligent Database Assistant", *IEEE Expert 1,2* (Summer): 65-79.

Kerschberg, L., ed., 1986. *Expert Database Systems: Proceedings from the First International Workshop.* Menlo Park, Calif.: Benjamin/Cummings.

Law, K.H., T.F. Zimmie, and D.R. Chapman, 1986. "An Expert System for Inactive Hazardous Waste Site Characterization", in *Expert Systems in Civil Engineering,* edited by C.L. Kostem and M.L. Maher, New York: American Society of Civil Engineers.

Levitt, R.E., and J.C. Kunz, 1985. "Using Knowledge of Construction and Project Management for Automated Schedule Updating", *Project Management Journal* 16, 5 (December): 57-76.

Maser, K.R., 1986. "Applications of Automated Interpretation to Sensor Data", in *Expert Systems in Civil Engineering,* edited by C.L. Kostem and M.L. Maher. New York: American Society of Civil Engineers.

Prerau, D.S., 1985. "Selection of an Appropriate Domain", *AI Magazine* 6,2 (Summer): 26-30.

Rauch-Hinden, W.B., 1985. *Artificial Intelligence in Business, Science and Industry, Volume II: Applications",* Englewood Cliffs, N.J.: Prentice-Hall.

Ritchie, S.G., and J.P. Mahoney, 1986. "A Surface Condition Expert System for Pavement Rehabilitation", paper presented at the American Society of Civil Engineers convention, Seattle, April 9.

Tanic, E., 1986. "Urban Planning and Artificial Intelligence: The URBYS System", *Computers, Environment and Urban Systems* 10, 3/4: 135-146.

Waterman, D.A., 1985. *A Guide to Expert Systems,* Reading, Mass.: Addison-Wesley.

Williams, G.J., P.M. Nanninga, and J.R. Davis, 1986. "GEM: A Microcomputer Based Expert System for Geographic Domain", in *Proceedings of the Sixth International Conference on Expert Systems and Their Applications,* Vol. 1. Proceedings of a conference held in Avignon, France, April 28-30.

Zozaya-Gorostiza, C., and C.T. Hendrickson, 1986. "An Expert System for Traffic Signal Setting Assistance", Paper presented at the American Society of Civil Engineers convention, Seattle, April 9.

CHAPTER 2

Expert Systems Technology[1]

Lyna L. Wiggins, Sang-Yun Han,
Catherine D. Perman and Yi-Chin Lee

There are currently two principal approaches to developing an expert system for a particular application. The first approach is to use a programming language and write original code from scratch for the system. The second approach is to rely on one of the tools developed specifically to aid in the construction of expert systems. These tools are generally called *shells*. You will find that the urban planning applications of expert systems described in this book have made use of both approaches.

When the use of a programming language is the approach selected, nearly any higher-level programming language can be used, although some choices have been more popular than others. For example, in reviewing approximately 136 expert system applications, Harmon identifies 11 which are programmed in some dialect of Lisp, 3 in Prolog, 2 in C, and 1 each in Basic, Fortran, Pascal, and Cobol (Harmon, 1986). Notice that some of these systems are written in "traditional" languages such as Fortran and Pascal, which are not usually associated with artificial intelligence applications. The site selection application developed by Findikaki in Chapter 11 in this book makes use of Pascal. The zoning application described by Davis and Grant in Chapter 4 is written in Fortran.

Lisp, Prolog and Smalltalk are often called AI languages because of their particular characteristics. Lisp is a language based on *list processing*. In Lisp, the data structure consists of *atoms* (an element that can not be subdivided) and

---

1 Portions of this chapter were published in "A Planners Review of PC Software and Technology," Edited by Richard E. Klosterman, *Planning Advisory Service Reports 414/415*, American Planning Association, Chicago, IL, 1988.

"lists" of atoms. Lisp is the second oldest (Fortran is the first) programming language in use today (Hu, 1987). Prolog has been a popular choice in Japan, is based on *logic programming*, and has a syntax which relies on statements about objects and their relationships. In Chapter 5 in this book, Richard Brail describes a prototype system developed in Prolog. Brail also discusses some of the advantages and disadvantages of this language for implementing his system. Smalltalk is an *object-oriented* programming language, and uses *objects* (basic elements), *classes* (sets of objects), and *messages* (a request for an object to perform an operation). Object-oriented programming is praised for its ability to allow new code to be added to a program without modifying existing code.

By far the most popular approach to developing expert system applications today is the use of a shell. The idea of a shell is associated with the development of the early expert system called MYCIN (Buchanan and Shortliffe 1984). This early system was developed to diagnose infectious blood diseases. The important innovation was the separation of the *knowledge base* portion of the system from the *inference engine*. This concept of separation led to the development of EMYCIN (Empty MYCIN). With a shell, the programmer is free from writing original code to implement the inference engine portion of the system. The commercial shells also generally supply a higher-level language making it easier to write rules. The majority of the expert systems applications in urban planning described in this book were implemented using commercial shells. These shells include: KES (Knowledge Engineering System) in Chapter 3 by Wright; GURU in Chapter 6 by Southworth; Personal Consultant Plus in Chapter 8 by Suh, Kim and Kim, in Chapter 9 by Han and Kim, and in Chapter 13 by Kim and Adams; INSIGHT 2+ in Chapter 10 by Rouhani and Kangari; and NEXPERT in Chapter 13 by Lee and Wiggins.

Fiegenbaum, McCorduck and Nii note that the trend in recent years has increasingly been toward users acting as their own knowledge engineers (Fiegenbaum, McCorduck and Nii, 1988). This is largely a function of the availability of these shells, particularly those for microcomputer platforms, from third-party software vendors. Having tools that are increasingly easy for a "domain expert" to use, without the aid of a specialized programmer, is likely to mean that the dominant mode of system development becomes the expert creating their own system. Just as planners have adopted other desktop packages, such as spreadsheets, for their own application development, we will see expanded use of expert system shells for planning-domain problems in the next few years. In this chapter we will examine five shells that have been implemented for a microcomputer platform: (1) INSIGHT 2+ by Level Five Research, Inc.; (2) NEXPERT Object 1.0 by Neutron Data Inc.; (3) GURU by Micro Database Systems Inc.; (4) Personal Consultant Plus by Texas Instrument, Inc.; and (5) KEE by Intellicorp, Inc. In each case, we wish to be observant about how the shell handles their most important functions—the editing function, the control function and the explanation of chains of reasoning. Each description will conclude with a summary of the strengths and weaknesses of the shell.

Because much of the expert systems literature is particularly full of specialized jargon, in this chapter we take care to define and introduce many useful terms and concepts as we describe each of the shells. Since we also describe the snytax and form of some rules, as they are used in the various shells, this chapter also acts as a first tutorial in writing production rules in expert systems. A familiarity with this vocabulary and rule forms will be helpful in reading about the expert systems described in this book.

Categorization of the Five Shells

Harmon has suggested a 8-category typology of expert system tools (Harmon, 1986). These include 4 major categories: (1) hybrid tools, providing complex development environments, including "multiple paradigms" and graphics; (2) rule-based tools, providing simpler capability to create IF...THEN rule structures; (3) inductive tools, allowing the generation of rules from examples; and (4) domain specific tools, providing specialized tools restricted to a particular domain. In addition, Harmon divides the hybrid tools into "large" and "mid-size," according to the hardware platform required to run the shell. The large shells require mainframes, Lisp machines (workstations designed specifically to process Lisp), or Unix workstations. The mid-size shells may be run on personal computers (PCs), and are relatively new to the market. Harmon also divides the inductive systems into "large" and "small" (PC-based), and the rule-based tools by size into "large," "mid-size," and "small." The differentiation of the small from the mid-sized categories for the rule-based tools is used to describe the PC-based shells which do not provide context trees, confidence factors, and some other editing features.

Using this 8-category typology, the 5 shells reviewed in this chapter are classified by Harmon as follows: (1) INSIGHT 2+ (small rule-based); (2) NEXPERT (mid-sized rule-based); (3) GURU (mid-size rule-based); (4) Personal Consultant Plus (mid-size hybrid); and (5) KEE (large hybrid). Using this classification, it may be seen that the 5 shells represent a range of the existing commercial shells. KEE, classified by Harmon as a large system, was originally developed for workstation platforms. In 1988, it was ported to a 386 personal computer environment, and might now also be classified as mid-sized. These 5 shells are representative of the important classification classes of *hybrid* and *rule-based* systems of several sizes; they are not representative of the more specialized classifications of *inductive* and *domain-based* systems. In the following sections, we introduce these five shells roughly in the order of their overall complexity, from the simplest to the most hybrid.

INSIGHT 2+

Hardware configuration. The minimum hardware configuration for Insight 2+ is an IBM personal computer (PC, XT or AT, or compatible) with 256K RAM, two floppy disk drives, and a color or monochrome display. The recommended configuration is a minimum of 512K for full functionality.

The editing function. INSIGHT 2+ provides a high-level knowledge engineering language called PRL (Production Rule Language) for entering and organizing information into a knowledge base. The PRL is quite easy to learn, with straightforward grammatical rules. The stated goal of the vendor is to provide a system that can be used by the expert, without the necessity of a programmer to serve as a translator. The PRL source file can be created with a text editor of your choice, or it can be created with the Insight 2+ text editor (which is similar to Wordstar). The user must structure the knowledge base into a series of goals and their supporting rules. An example might be:

```
RULE 6.0

    IF land use IS residential
    AND zoning is R1
    THEN ordinance is 10.1 (CF = 100)
```

Notice here that the rule has been named `6.0`. The principal data construct used in the final line is the object-attribute-value (OAV) statement. Such OAV triplets, and simpler attribute-value (AV) pairs, are a common method of representing factual knowledge in expert system shells. In this simple example, we are working with a single object, a particular development site for which we wish to classify the appropriate section of the zoning ordinance. In OAV triplets, the "object" is an actual entity (e.g., the particular development site); the "attribute" is a property associated with the object (e.g., the land-use classification, the zoning classification), and the "value" of an attribute gives the nature of the attribute in a specific situation (e.g., residential and R1). In this simple example, the user of the system would have been asked to provide the values "residential" and "R1" during the query process, and these values would be stored in working memory. The process of determining these specific values is called "instantiation." The conclusion here is that Section 10.1 of the zoning ordinance is applicable for this particular development site, and this conclusion is reached with a Confidence Factor (CF) of 100, which implies absolute certainty.

Insight 2+ makes extensive use of CFs in both queries and conclusions. When a particular line of reasoning is being pursued, Insight 2+ chooses first those rules that might achieve the highest degree of confidence. The user of the system may also be asked to enter his or her level of confidence in a particular answer, and the system developer may enter minimum threshold levels (e.g., the minimum acceptable level for a positive response) for particular queries. The user enters a confidence level for queries through a bar graph on the screen by using the left and right arrow keys to select a level shown on the graph.

The PRL also supports string data (e.g. John), numeric information (e.g., today's high is > 90), and factual information (true, false). Within the PRL, other programs may be called and Pascal programs may be executed. For example, the system developer might want to execute a program in Pascal which does a complicated calculation and pass the results back to the knowledge base. Another example might be the need to obtain information from a data base, or return updated information to a data base. Insight 2+ offers a particularly easy interface to dBase.

Often, the user of a system may require additional information to answer a particular query. Insight 2+ provides an easy mechanism for the system developer to add additional information. This information may then be displayed upon request by the user. There is no limit to the number of lines of this expanded text. Text may also be shown automatically without the user requesting the information.

After a knowledge base has been written in the PRL, it must be compiled. During compiling, error messages are given and must be corrected. After a successful compilation, the system may then be run by the system user in what is called *run time*. Each time a change in the PRL is desired, the system must be recompiled.

The control mechanism. One of the strengths of the expert system shells is that system developers are free from the programming that tells the computer "what to do" and can spend most of their time describing the relationships between facts, events and things in the production rules. The shell itself deals with the actual control of the program. Insight 2+ makes use of goal sets, which are mutually exclusive groupings of goals. Run-time users of the system first select the group of goals that are the most appropriate for the problem at hand. The control mechanism then used is *backward chaining*. Backward chaining systems are sometimes called *goal-directed* systems. The possible outcomes are known, generally small in number, and the inference engine works backwards through the rule base to choose an answer. Using backward chaining, the system attempts to find out if the goal rule is correct by going back to the "if" clauses of the goal rule and determining if they are correct. If these clauses are correct, the system continues to examine other confirming rules backward through the system.

The line of reasoning. INSIGHT 2+ provides a variety of displays that allow the user to examine the line of reasoning. The conclusion display gives a summary of all the intermediate and final conclusions from a session. The user may ask to show the current rule, and is then given a display of the PRL source code for that specific rule. The user may also ask for a display of all the facts (responses to queries) that were entered during the session. These facts may be changed at this point, providing an easy way to ask "what if" questions. A final report function gives the entire line of reasoning used, in the order they were used during the session.

Advantages and disadvantages. INSIGHT 2+ has strong advantages, particularly for novice expert systems developers. This shell has well written documentation, is easy for the system developer to learn, and provides good error trapping during compilation of the PRL. In its run-time mode, the user is given a menu structure and help functions. The run-time user would need no prior computing experience to operate the system. It has minimal hardware requirements. The CALL and RETURN functions provide straight-forward linkages to other micro packages. INSIGHT 2+ makes it extremely easy for the system developer to add supporting text information for the user. The report functions provide the major information about the session—facts, rules, conclusions, and the chain of reasoning—in terms that are like normal language.

INSIGHT 2+ is text-oriented and does not support graphics. Although each of the screens and reports is clear, they are not particularly interesting visually. An example of a typical screen is given in Figure 2.1. The most entertaining feature for the run-time user is moving the arrow keys back and forth for queries that contain confidence factors. Answering query after query in this form becomes tedious after a few hours, and this might be a problem for a complicated system or for one that the user must access frequently. Although using a compiled knowledge base speeds the response time for the run-time user, it requires the system developer to recompile each time a rule is added or changed. A final major disadvantage of the software is the restriction to backward chaining only, as we will see when we look at the other two system shells. To be successful, systems which only allow backward chaining must be used for problems where the number of goals is reasonably small and the possible outcomes are known.

NEXPERT Object

Hardware configuration. NEXPERT Object 1.0 runs on the Macintosh Plus, SE and Macintosh II with 1 megabyte of memory or more. It takes advantage of the color display and multiple windows of the MAC II.

The editing function. NEXPERT takes full advantage of the Macintosh interface (e.g., pop-up windows, dialogue boxes) for editing the rules in a knowledge base. Rules are created in a special window environment. Three types of data are allowed: Boolean (true,false), string and numeric. The operators (is, is not, equal, not equal, etc.) are entered into the rule window environment by clicking on a pop-up window. NEXPERT does error checking in real time as you enter your rules. If an error is indicated, a *check* function highlights the source of the snytactically incorrect rule. For later editing, the system developer can choose a pop-up menu that provides a list of rules by name. Selecting one, you may then add, copy, modify or delete.

Making use of the Macintosh interface, the NEXPERT software provides the system developer with the option of providing both text and graphic information to the system user. For example, a documentation icon may be placed over a particular hypothesis in the network. When the user clicks on the icon, an image created from MacPaint or another graphics package is displayed. This provides the user with additional graphic information about the hypothesis.

CORA Expert System Main Calling Routine

Select all waste types that apply to the site in question
F1 for additional selections
F5 for help

=== Uncontaminated unsaturated soils

Homogeneous contaminated unsaturated soils

Hot spots {Unsat mtl around leaky tanks or drums}

Contaminated saturated soils {Groundwater}

Mixed debris

Stream covering contaminated sediments

Lagoon or surface impoundment

After making your selections press F4 for DONE

1 PAGE   2 UNKN   3 STRT   4 DONE   5 EXPL   6 S/R   8 MENU   9 HELP

Figure 2.1. Example screen from an Insight 2+ application. Notice that it is very text oriented. The run-time user would use the arrow keys and the return key to make their selections. This screen is from a system called CORA, developed for the U.S. Environmental Protection Agency by CH2MHill.

The control mechanism. NEXPERT provides *network navigation* to the system developer and system user. This is an automatically generated web which is displayed graphically as a tree. The user may select a "chunk" of the network, and then scroll backwards and forwards through the tree. A typical tree is given in Figure 2.2. Notice the graphic links between rules. NEXPERT supports both backward and *forward chaining*. Forward chaining operates by comparing the facts available in working memory to the premises of the rules. When a rule "succeeds," its conclusion(s) is then placed in working memory. The analogy is often made to a generalist approach to problem solving, where one first inquires in a general way about aspects of a problem (Harmon and King, 1985).

During a specific session with NEXPERT, the user would have access to several windows which provide control of the session. The *session control* window allows the user to specify the current task and enter information to queries. At any time, the user can use an *encyclopedia menu* to see and edit a full list of hypotheses and current data. Every rule and entity in the rule base is regarded

```
Yes Module_1_Completed
Yes Module_2_Completed
Yes Module_3_Completed
Yes Module_4_Completed
Yes Module_5_Completed
Yes Module_6_Completed ─ r.5 ─────────────── MEDIATOR_Completed
Yes Module_7_Completed
Yes Module_8_Completed
Yes Module_9_Completed
Yes Module_10_Completed
   =>Show ThatsAllFolks
```

```
Yes project_in_range
No More_Sites
                          r.5 ─────────────── Module_1_Completed
Yes Site_Information_Acquired
Yes Capacity_Known
```

```
No More_participants
                            r.1 ─────────────── Module_2_Completed
Yes Participant_Information_Acquired
```

```
Yes Num_Of_Env_Known
Yes Num_Of_Gov_Known
Yes Num_Of_Res_Known
Yes Num_Of_Dev_Known ─ r.6 ─────────────── Module_3_Completed
Yes Num_Of_Zoning_Known
Yes Num_Of_COE_Known
Yes Num_Of_CEC_Known
```

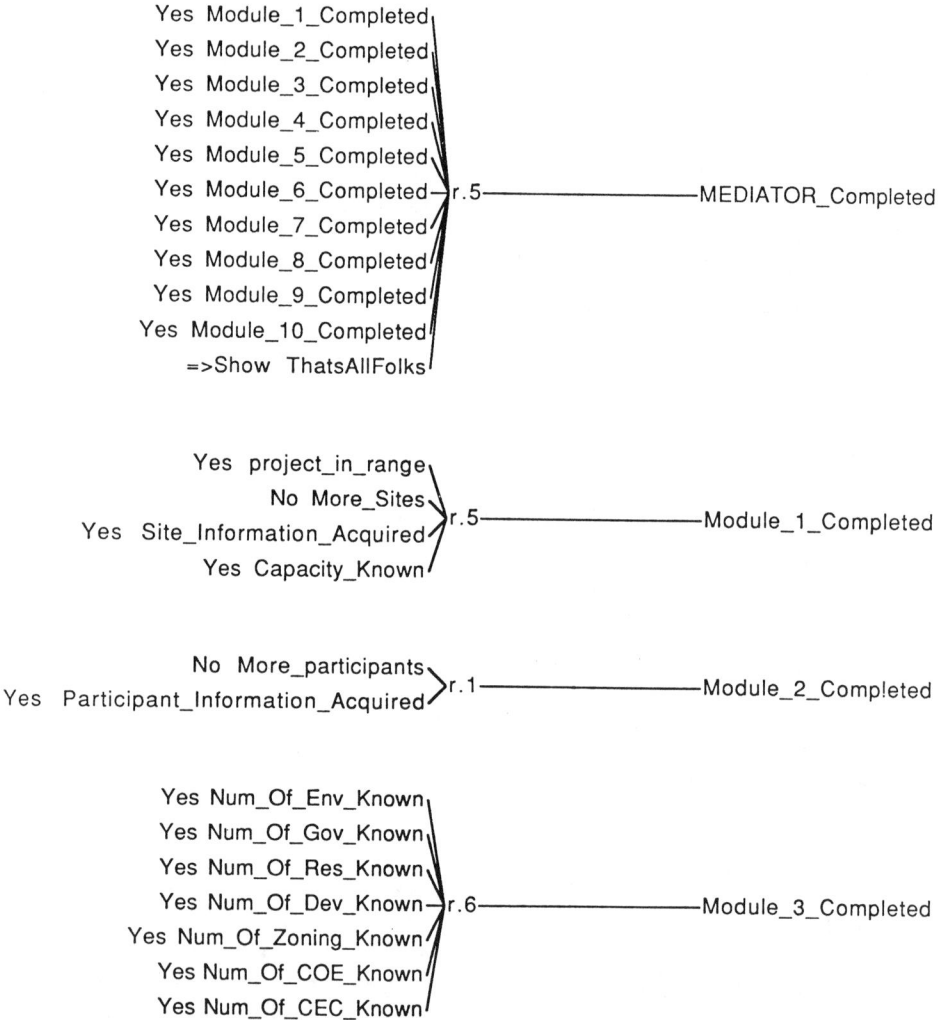

Figure 2.2. Example rule tree from a NEXPERT application. During system use, various icons and highlighting would show exactly which rules had been evaluated. This screen is from a system called MEDIATOR, developed at Stanford University by Yi-Chin Lee.

as an object by NEXPERT. NEXPERT supports a version of frames and allows *inheritance*. These concepts will be discussed in more detail in the section describing KEE.

NEXPERT also supports links to Excel (a spreadsheet package for the Macintosh) and to programs written in C.

The line of reasoning. Other windows are available to the user to aid their understanding of the line of reasoning. The transcript window monitors the session, the conclusions window keeps track of all conclusions reached, and the current hypothesis window indicates the active question under investigation. Of particular importance in understanding the line of reasoning is the graphic representation of the tree. The tree may be "inspected" by the user through the use of icons. For example, at each tree node, a check mark indicates that a condition is true, a highlighted check mark indicates false, a question mark indicates a currently unknown condition, and a target indicates that the condition is currently being investigated.

Advantages and disadvantages. NEXPERT benefits from the Macintosh interface. It has some of the best editors for modifying rules and objects we have seen. By using cut-and-paste, errors during rule entry are minimized. For example, the system developer may pull down a menu containing all previous objects. Clicking on one of the objects will paste that object to the current cell being entered. This feature not only prevents typos, but also serves as a reminder of previous objects. All new or modified rules are compiled and run in real time. Without having to recompile after each change, the system developer may speedily test out new and modified rules.

The most obvious advantage of NEXPERT is the graphic representation of the rule network. One can focus on specific nodes of the tree(s) or, if desired, the whole tree, to examine the current status of the rules. This is an ingenious evolution from the HOW and WHY questions used in earlier systems such as MYCIN.

NEXPERT is a complicated, and potentially powerful software package. With such a complicated system, a well written user manual is a necessity. This is unfortunately not the case with NEXPERT. The manual is poorly written. Many of the examples are unclear and too primitive to be useful in explaining concepts. Often the only way to discover how to execute particular operations is to examine the sample programs. The manual does not aid the user in creating even the simplest applications. Trivial mathematical calculations are extremely awkward in NEXPERT. Although there is a link to Excel, the user must quit NEXPERT, open specific Excel files, let the calculation run, then quit Excel, come back, and open up another NEXPERT knowledge base that retrieves and makes use of the calculation. It's far from a simple manipulation. To run external programs like C or Pascal, the user must have the required compilers and utilities. Again, the manual did not provide useful information for actually implementing these links.

Another shortcoming is the inability of Nexpert to isolate a dynamic object for inspection. A dynamic object is one created during the process of inference. For example, in one system we developed, we wished to interactively enter the specific names of participants (having properties provided by the class

participant) and give them an identification number. We also wanted to create a
list of the names of these participants, make them into a menu, and then allow
the user to make multiple selections. Neither of these should be a difficult task,
but they were not possible within NEXPERT. The attributes of dynamic objects
could not be retrieved, and multiple selection from a menu is not available.
Finally, the conclusions reached during a session need to be returned to the user.
Here NEXPERT allows only "canned text" (e.g., prewritten conclusions), rather
than text which is generated during the session (*floating text*).

GURU (Micro Database Systems Inc.)

Hardware configuration. The minimum hardware requirement for GURU is an
IBM-PC, XT, AT or compatible with 640K RAM, monochrome display, and
five megabytes of hard disk space. Using graphics during a consultation
requires an EGA card and monitor. For hard-copy output of graphics, a graphic
printing device is needed. GURU supports the HP 7475 plotter and the IBM
Graphics Printer (or compatible) for printing graphics. An optional input device
is a mouse, and all of the major brands are supported.

The editing function. GURU provides a programming environment for both
conventional procedural programming and production rule programming. It
provides a built-in text editor. Developers can write a mixture of IF-THEN production rules and conventional procedural (IF-THEN-ELSE, DO-WHILE) programs. Therefore, in structuring the knowledge base, system developers have
two choices: 1) use a pure production rule format which requires designation of
goal variables and a series of IF-THEN rules; and/or 2) in addition to production
rules, write necessary algorithmic programs and let the production rules call
algorithmic procedures whenever necessary.

   For instance, in GURU the production rules may be written simply in
PREMISE-ACTION syntax:

```
GOAL:   advice

RULE:   rule1

      IF:   erosion_potential = high
            AND flood_potential = high
    THEN:   environmental_suitability = low

RULE:   rule2

      IF: soil = Type_A
    THEN: erosion_potential = high AND ....

RULE:   rule3

      IF: environmental_suitability = low AND ....
    THEN: advice = consider_other_sites
```

Besides IF-THEN parts of a rule, users can write additional information which includes:

```
REASON:    to provide expert system users with the
           reason for the rule in response to WHY command
PRIORITY:  to indicate the rule's priority using integer
           values from 1 to 100
COMMENT:   to write comments (up to 255 characters) to
           the expert system developer
     cf:   to indicate confidence factors (0 through 100)
```

Thus, more complete rules might appear as:

```
RULE:    rulename

    IF:    land use = residential
  THEN:    ordinance = 10

REASON:    All residential land uses fall under
           section 10 of the zoning ordinance.
COMMENT:   Part of the classification portion
           of the system.
PRIORITY:  20
```

Additional attributes can be attached to a rule to indicate how a variable's values can be obtained (FIND), the relative costs of firing rules (COST), the number of times that a rule can be fired during one consultation (CAP), and so on.

If the format of the pure production rules is not sufficient to develop a system, algorithmic procedures can be added in various ways. The simplest way of integrating algorithmic procedures is to add a attribute called FIND into a rule set to indicate how a certain variable's values are to be obtained. These value assignments can be: 1) a simple input command to obtain the necessary information from the user: or 2) an extensive procedure file which calculates the necessary values. To illustrate the first type of use, consider the following example given in GURU syntax:

```
FIND:    @ 10,10 OUTPUT "What is the parcel number?"
      looptest = false
      WHILE looptest = false DO
          @ 10,30 INPUT parcel INTEGER USING "uuuuu"
          IF parcel <= 0 or parcel > 2000 THEN
          OUTPUT "Invalid data. Enter again"
          ELSE
          looptest = true
          ENDIF
      ENDWHILE
```

This is a simple example of incorporating conventional procedures into a rule set. Very complex algorithmic procedures can be written using IF-THEN-ELSE and DO-WHILE looping and mathematical functions including root, sine, etc. The support of such algorithmic programming by GURU can be useful for many planning applications of expert systems in which symbolic data processing alone is not sufficient. For instance, the numerical procedures of calculating internal rates of return or present values can be easily written and integrated into the expert system.

Once a knowledge base has been written in either the pure production rule format or in a mixture of rules and algorithmic procedures, it needs to be compiled for a test consultation. During compiling, detected errors are reported along with possible explanations for the error.

The control mechanism. To help the developers of an expert system grasp the overall structure of the knowledge base, GURU provides a utility to create a graphical representation of the rules and their associated variables for a rule set. *Tracing* is another utility by which developers can trace rules fired during a consultation tation so that developers can check the logic of the system. Utilities such as these are useful during the process of building a prototype system, refining it with the aid of the source expert, and expanding it into more complete versions.

The type and flexibility of control mechanism provided for the end-user of the expert system depends on how the developers (and programmers) design the system. Unlike many other expert system shells, GURU does not provide built-in input and output facilities for final applications. This may provide more flexibility to the system developer, but it can also increase the overall development time. Compared to third generation languages such as PASCAL and FORTRAN, however, it is relatively easy to develop efficient input/output facilities using GURU commands. The input facilities, including menu structure, type of screen windows, and use of function keys, can be programmed using GURU's internal commands. Customized reports can be also designed using GURU commands. A typical screen from a GURU expert system in given in Figure 2.3.

As for the reasoning mechanism, GURU supports backward, forward, and mixed chaining. Backward chaining causes GURU to identify candidate rules for a goal variable based upon their conclusions (THEN parts). GURU processes one or more of these candidate rules to establish a value for the goal variable. Forward chaining causes GURU to reason in a forward direction (from the premises of rules to their conclusions), in order to test the effect of a rule set on the value of a variable. In mixed chaining, the direction of the reasoning process can be changed dynamically. For instance, in a consultation using backward chaining, a different rule set can be consulted using forward chaining. Users or developers can select the desired reasoning mechanism by using GURU commands: (1) CONSULT TO TEST for forward chaining; and (2) CONSULT TO SEEK for backward chaining. Mixed chaining may be initiated by setting an environment variable called E.MIX to R(rigorous).

```
┌────────────────── GURU RULE SET MANAGER ──────────────────┐
│  ──── Rule Set:londuse.rss ──────────────── Goal:occeptobility ──┘
├──── Knowledge Tree ─────────────────────────────────────┐
│                                                           │
│  Proposal                                                 │
│      │                                                    │
│      └── Evaluation                                       │
│              │                                            │
│              └── Compatibility                            │
│                      │                                    │
│                      ├── Industrial                       │
│                      │                                    │
│                      └── Residential                      │
│                                                           │
│                                                           │
│                                                           │
│                                                           │
└───────────────────────────────────────────────────────────┘
```

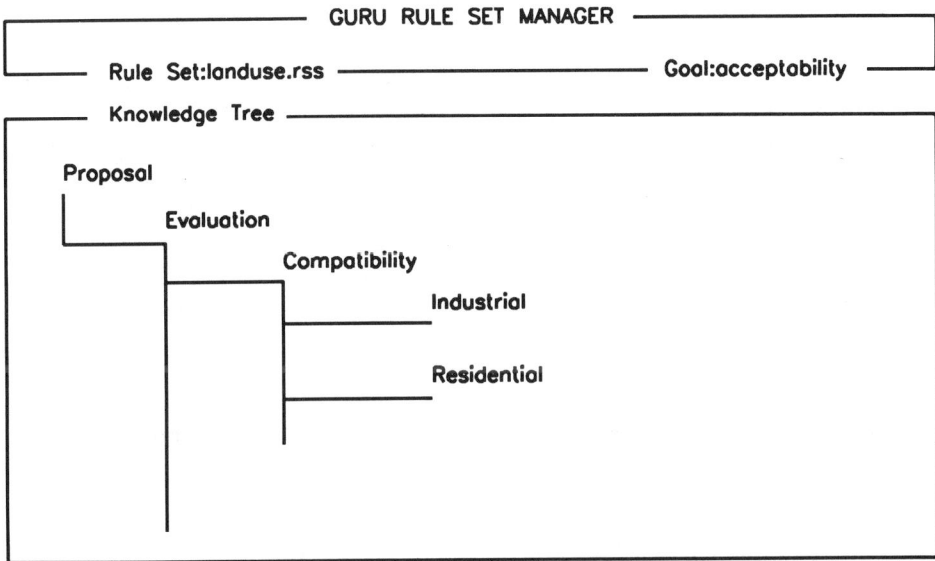

Figure 2.3. Example screen from a GURU application. The structure of the knowledge base is displayed.

Environment variables control some aspect of the interaction between users (or system developers) and the system. For instance, the environmental variable E.BAGC controls the background color of the screen and the variable E.CAP determines the number of times a rule can be fired during a consultation.

The line of reasoning. As discussed above, GURU provides a flexible reasoning mechanism. The system developer may try different chaining methods and see how each method navigates through the knowledge base. By checking the printed output of trace results, the developer may select a reasoning method which is best suited for a given problem. The system developer may also design the system in a way that allows the users to select their preferred reasoning methods.

Advantages and disadvantages. Although only the expert system aspects of GURU have been explored here, GURU actually provides an environment ment for developing an *integrated* information system. In addition to the expert system development tool, GURU provides a spreadsheet, database management, natural language interface, text processing, and remote communication capabilities. One of the current limitations of personal computers is imposed by the popular disk operation system (DOS) in that it does not allow multi-tasking (i.e. running multiple programs simultaneously). GURU has a clear advantage in allowing the use of several different programs within one environment. The database management system inside GURU can be effectively used to store a lot of facts, eliminating the need for writing many extra production rules.

The problem with this integrated tool, however, is that the developer must put additional effort into learning all of the various tools available in GURU. The spreadsheet program within GURU looks very different from many standard spreadsheet programs, including the well known Lotus 1-2-3. The database management system in GURU is also relatively poor compared to many popular database programs such as dBase or Rbase.

In short, GURU may not be suitable for building a *prototype* system which requires a fast development cycle. For prototyping, other expert system shells which provide built-in user interfaces and require less programming knowledge may be preferable. GURU may be more suitable for building a final application. Because GURU is written in the programming language C, it works very fast and requires less computer memory compared to the shells written in Lisp. In addition, the support of procedural programming, database management, and graphic routines makes GURU a suitable tool for building final applications of expert systems.

Personal Consultant Plus (Texas Instruments, Inc)

Hardware configuration. The minimum hardware requirements for Personal Consultant Plus (PC-Plus) is an IBM-PC, XT, AT or compatible with 512K RAM, monochrome display, and 1.5 megabytes of hard disk space. Because PC-Plus is written in Lisp, which easily consumes computer memory, and because expert systems developed in urban planning are likely to have a large knowledge base with more than 200 rules, IBM- AT class machines with extra memory expansion cards are strongly recommended. Extended or expanded memory of up to 2 megabytes is supported. Using graphics in consultation requires an EGA card and EGA monitor.

The editing function. PC-Plus provides a built-in text editor for writing rules (see Figure 2.4). In developing a knowledge base in PC-Plus, it is important to understand three different concepts: (1) parameter; (2) rule; and (3) frame. Parameters are specific facts or pieces of information (e.g., population size and density). Rules are logical statements written in a rule language provided by PC-Plus, and they describe logical relationships between parameter values. Frames serve a general organizational purpose. A frame defines a real or abstract problem area in the knowledge base. (In the next section of this chapter, on KEE, we will see a system that makes extensive use of frame structures.)

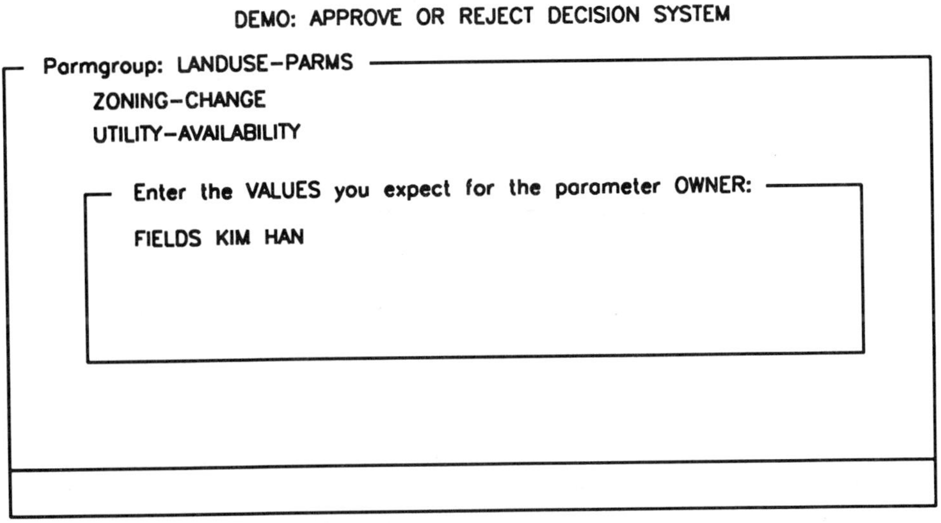

DEMO: APPROVE OR REJECT DECISION SYSTEM

┌─ Parmgroup: LANDUSE-PARMS ─────────────────────────────────────┐
│   ZONING-CHANGE   │
│   UTILITY-AVAILABILITY   │
│   │
│   ┌─ Enter the VALUES you expect for the parameter OWNER: ─┐│
│   │ FIELDS KIM HAN   ││
│   │   ││
│   │   ││
│   │   ││
│   │   ││
│   └──┘│
│   │
│   │
└───┘

Figure 2.4. Example screen from a Personal Consultant Plus application. A sample of the menu system for knowledge base development is shown.

To illustrate the use of these concepts in developing a knowledge base using PC-Plus, consider a simple example of developing an expert system for landfill site selection. Parameters may include such things as selected site (SELECTED_SITE), area of the site (AREA_OF_SITE), depth to groundwater on the site (DEPTH_TO_GROUNDWATER), land uses on adjacent properties (SURROUNDING_LANDUSE), and so on. When defining parameters, the system developer needs to declare certain properties of a parameter, including such properties as the type of valid parameter values and how these values can be obtained. Basically, the system developer can specify one of three optional sources of parameter values: 1) end-users; 2) external database; and 3) rules in the knowledge base. For instance, the value of the parameter SELECTED_SITE may be obtained from the knowledge base after trying necessary rules, whereas the value of the parameter AREA_OF_SITE may be obtained directly from the user during a consultation.

Once necessary parameters and their properties are defined, IF-THEN production rules can be developed to show the logical relationships between parameters. Rules can be written in either English-like syntax (called ARL:

Abbreviated Rule Language) or in Lisp syntax. If ARL syntax is preferred, the developer simply answers the prompts provided by PC-Plus to enter the IF part, THEN part, and EXPLANATION part for WHY questions. Then PC-Plus converts the ARL form to the correct Lisp form. For instance, the developer may write the following rule in ARL syntax:

```
IF  (occupation = planner OR occupation = professor)
 AND (hobby = writing OR hobby = drawing)
THEN  intelligence = high
```

The above rule will be written in Lisp syntax as below:

```
PREMISE :: ($AND
           ($OR
           (SAME FRAME OCCUPATION PLANNER)
           (SAME FRAME OCCUPATION PROFESSOR))
           ($OR
           (SAME FRAME HOBBY WRITING)
           (SAME FRAME HOBBY DRAWING)))
ACTION ::  (DO-ALL
           (CONCLUDE FRAME INTELLIGENCE HIGH))
```

In PC-Plus, it is useful to organize rules into several frames when the number of rules is very large and/or when the knowledge base consists of several sub-domains. For instance, an expert system for landfill site selection may consist of several sub-domains (e.g., construction engineering, water pollution, and legal domains). Each frame can contain rules and parameters relevant to that particular frame. In aiding a programmer in developing parameters, rules, and frames, PC-Plus provides good editing and debugging facilities. It detects and reports missing parameters and multiple definitions of the same parameter.

The control mechanism. PC-Plus primarily uses backward chaining. As described earlier, this reasoning method works from a goal to supporting facts. One of the good features of PC-Plus is that the developer can specify multiple goals at one time so that the system tries to satisfy multiple goals until it finally succeeds or fails. This can be useful for many planning applications of expert systems. For instance, an expert system for landfill site selection may be designed to seek multiple goals: one goal of finding a suitable site from the perspective of environmentalists and another goal of finding a suitable site from a legal perspective. Assuming that the environmentalists and legal experts have different problem solving approaches to landfill site selection, and that their expertise and heuristics are implementable into an expert system, the use of multiple goal specification will give very interesting results.

To help the system developer grasp and control the overall structure of the knowledge base, PC-Plus provides a utility for depicting the hierarchical structure of the knowledge base in graphic form. To aid interactions with system users, PC-Plus also allows the incorporation of graphics as a part of a question to the user or as a part of an answer by the system. Any third-party graphics programs may be used to create graphics (maps or diagrams) which can be imported into PC-Plus using provided utilities. For a more interactive use of graphics, a separate graphics program called Personal Consultant Image may be used. (This program is provided at extra cost.)

The line of reasoning. As explained earlier, PC-Plus primarily uses the backward chaining reasoning method. However, the system developer can introduce forward chaining whenever desired by simply assigning an additional attribute called ANTECEDENT to a rule. This can be useful in a situation where some rules need to be fired regardless of the goal states. For instance, in the example of landfill site selection, the rules which determine the type of waste material may need to be fired at the beginning of each consultation. By giving the ANTECEDENT property to this rule, it will be fired even when it is not necessary to the backward chaining process.

PC-Plus offers additional methods for more advanced reasoning. One example includes assigning a property called DO-BEFORE. By assigning DO-BEFORE to a rule, the system developer can specify the order of the firing of rules. For example, when two rules solve the same problem while one rule saves wetland and another rule saves cost, the system developer can specify which rule has first priority. Other available methods in PC-Plus for advanced control of reasoning includes UTILITY, a property by which the developer assigns relative utility to a rule, and *meta-rules,* by which the developer can write high-level knowledge that determines the most useful rules during a consultation.

Advantages and disadvantages. PC-Plus may be suitable for both novice users as well as expert users of expert systems. For novice users, it provides good documentation, examples, tutorials, and menu-guided help facilities. Since input/output facilities are already provided in PC-Plus, novice users can just concentrate on developing a knowledge base. For expert users, PC-Plus provides very advanced features such as meta-rules. It also provides a good interface with dBase and Lotus 1-2-3 for effective database management and numeric computations. Some disadvantages may include the slow operation speed and high memory requirements which are intrinsic problems of many Lisp programs.

Knowledge Engineering Environment (KEE), Version 3.1

Hardware configuration. Over the last several years, Intellicorp, the vendor of KEE, has revised the shell to run on a variety of machines including Lisp machines (e.g., Symbolics 3600 series and Texas Instrument Explorers), engineering workstations (e.g., SUN 3 and Digital VAXstation III), and the 386-based personal computers (e.g., COMPAQ 386 and, in late 1988, IBM PS/2

Model 80. This review concentrates on KEE in the 386 environment. This implementation is a full development of KEE 3.1, with the same functionality as KEE running in other environments. Slight differences in appearance from machine to machine are dependent on certain computer monitor characteristics such as screen size and resolution. In the 386 environment, the required hardware includes a minimum of 10Mg of memory (suggested 12Mg), and minimum disk space of 100Mg (suggested 130Mg). Graphics cards required are EGA, VGA or compatible, and a Microsoft Bus Mouse is required for input.

The editing function. At the center of KEE is the object-oriented style of programming (Fikes and Kehler, 1985 and Stroustrup, 1988) in which the basic data structure is an object. The use of the word *object* refers to concepts and entities that describe an application problem. Objects can be abstract concepts, such as "all parents" or "all wastewater discharge permits," or specific entities, such as "my father" or "my wastewater discharge permit."

Frames are a type of object with associated characteristics or *attributes* that provide a place for storing information about the object. For example, the frames "all wastewater treatment permits" and "my wastewater discharge permit" would both have an attribute called "suspended solids limit." With KEE, frames are called *units* and frame attributes are called *slots*. Slots have characteristics called *facets* that provide extra information about the slot. For example, the slot suspended solids limit would have a facet called units that stores the kind of engineering units (e.g., mg/l) used to define the permit limitation. For each slot, KEE automatically provides facets that control the kinds and number of values permitted in the slot. In KEE, groups of frames are organized into files called *knowledge bases (KB's)*. A KEE application consists of one or more knowledge bases.

In order to complement the object-oriented style, KEE provides a construct, or a relation between units, called *inheritance* that allows the system developer to classify units into hierarchical taxonomies defined by inherited attributes. Abstract objects describing parent units are called *class units,* and the objects describing unique entities are called *member units* (e.g., as in a member of a class). For example, in an application about wastewater treatment, the unit called "all wastewater treatment plants" would be a parent unit of "my wastewater treatment plant." Through inheritance, my wastewater treatment plant unit will inherit the slot suspended solids limit from the "all wastewater treatment plants" unit. Inheritance provides a convenient way of organizing objects in a knowledge base and, more importantly, provides an efficient method for proliferating attributes for entire classes of objects. With inheritance, the system developer an write procedural code (e.g., Lisp code) and rules that can be applied to any member of a specific class.

Message passing is a feature of object-oriented programming that makes use of procedural code stored in an object attribute. With KEE, messages are called methods and are stored in slots. Method slots can be inherited like any other slot. This allows procedural code that describes the behavior of a class of objects to be inherited by a specific unit. With KEE message passing, the procedural code for a specific unit can be executed with a special call to the unit and the method slot, thereby dispensing with having to remember detailed information about the actual code in the slot. By using the object-oriented programming style, system developers can write rules and procedural code in Common Lisp that directly access and store information in specific objects without declaring additional variables. KEE provides a set of functions that directly access and store a unit's information. From the point of view of the system developer, KEE provides rule syntax that allows the system developer to write rules that access and store information directly in KEE units.

The KEE rule system (*Rulesystem3*) is based on a KEE logic language, called *TellandAsk*, that provides a production rule syntax with an English-like appearance. In addition, Rulesystem3 has a special syntax that provides an object-oriented interaction with KEE units. The Rulesystem3 provides complete pattern matching capabilities that allows the full use of variables. Whether intended for forward or backward chaining, all rules have the same syntax. In fact, the same rule can be used with different chaining mechanisms. In KEE, rules can be written with or without variables. The basic syntax follows the oft-used If premise then consequent format.

The control mechanism. KEE supports both backward and forward chaining. In addition, the system developer can initiate forward chaining from within backward chaining, since all rules have the same syntax. We will consider several examples of rules to illustrate how KEE structures its control mechanism. The first example of a KEE rule uses what KEE terms *unstructured facts* which do not refer to any of the application objects created by a system developer:

```
1)   (If (the treatment plant exceeds its permit limitation)
        then (the plant has an operating problem))
```

Note the Lisp-like use of parentheses as delimiters to define the premise, the consequent and the rule itself. If the system developer wants to refer directly to specific objects, the clause of the premise or consequent would take on the syntax *(the <slot> of <unit> is <value>)*. For example, here is the first example rewritten with the object oriented syntax:

```
2)   (If (the permit.limitation of treatment.plant is exceeded)
        then (the problem of treatment.plant is occurring))
```

With the second example, when the rule is fired by the inference mechanism, KEE checks the unit "treatment.plant" to see if the slot "permit.limitation" has the value "exceeded." If the value is exceeded, then the consequent will be asserted by adding the value "occurring" to the slot "problem" in the unit treatment.plant. With the complete pattern matching facilities, KEE rules can have variables, signified by a character symbol preceded by a question mark (e.g., ?X or ?Value). For the third example, the rule in example 2 could be rewritten with a variable in the <unit> location of the clauses:

```
3)  (If (the permit.limitation of ?treatment.plant is exceeded)
    then (the problem of ?treatment.plant is occurring))
```

In this example, the rule, when fired, will test the premise with whatever unit is bound to the variable. Variable bindings can be limited to members of specific class units by adding the following clause to the rule:

```
4)  (If (?treatment.plant is in class all.treatment plants)
    and (the permit.limitation of ?treatment.plant is exceeded)
    then (the problem of ?treatment.plant is occurring))
```

Now the rule of the final example is applicable to all members of the class all.treatment plants. Thus, it is an example of a rule that takes advantage of a frame-based representation.

In KEE, rules are implemented as objects. KEE has a special class of units called *Ruleclasses*. All rules implemented in KEE are descendants (using inheritance) of this class. The slots that are inherited from the Ruleclasses unit provide a link for a rule to the mechanisms of the Kee Rulesystem3. Typically, the system developer will create at least one rule class unit for the application which acts as an intermediate unit between the KEE ruleclass unit and the applications rules. Individual rules would be implemented as member units of the application's rule class.

Rules are fired by the use of one of two KEE functions: assert and query. Used as a Lisp function, assert initiates forward chaining and query initiates backward chaining. Each of these functions requires a rule class argument to start firing rules. By organizing the rule units into rule classes the system developer can control which rules could be fired with a specific assert or query. This can be used for meta-reasoning and for *pruning* the rule search for a specific assert or query.

The line of reasoning. KEE provides the system developer with two extensive interface toolkits. One set of tools, the KEE development interface, assists the system developer in designing, implementing and testing a system. Two examples of the KEE interface are given in Figures 2.5 and 2.6. As illustrated in these figures, the KEE development interface consists of multiple windows with bitmapped graphics that are used with a mouse and a keyboard. With KEE, the screens can be organized into desktops. The basic KEE desktop consists of KEE Icon, Global Menu Bar, Kee Output windows, a Lisp Listener window and Kee Typescript windows. Using the mouse on the Kee Icon and Global Menu brings up various menus for the KEE development interface. The Kee Output and Typescript windows display the various development interface graphics and reports. The Lisp Listener is for evaluating Lisp commands. In addition, KEE provides an editor on the desktop.

The system developer can use all of these tools to customize an interface with extensive graphics for the system user. The tools for the developer range from extremely easy to implement (with limited control over functionality and appearance), to more difficult to implement (with extensive control). Some of the easy tools include an ActiveImages package which provides a set of pre-

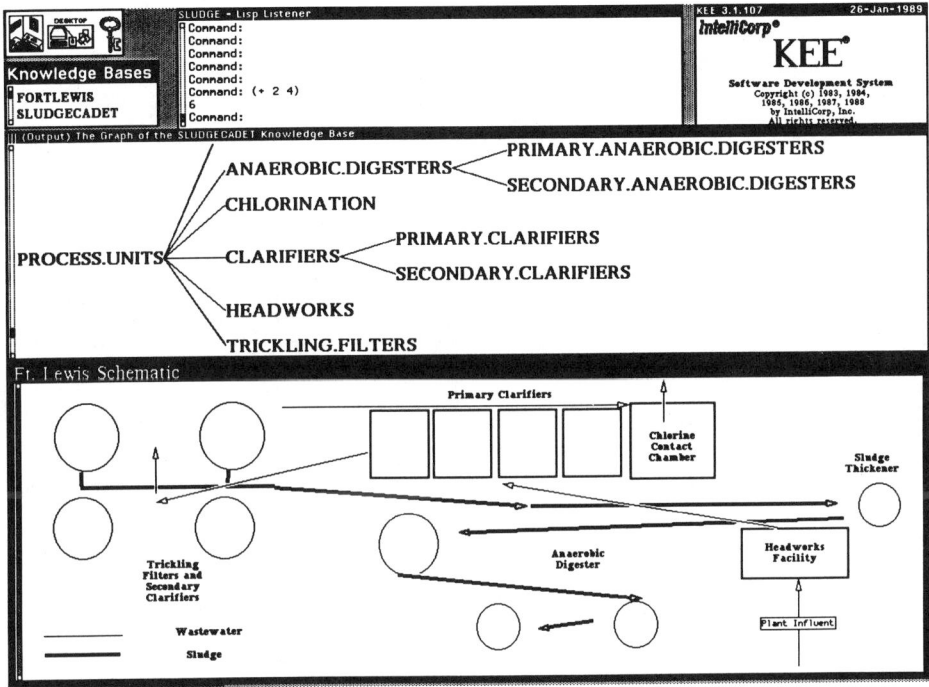

Figure 2.5. Example screen from a KEE application. Notice the use of multiple windows, including a rule tree and bitmapped graphics. This screen is from a system called SLUDGECADET, developed at Stanford University by Catherine Perman.

defined images with specific functions including value display, inputting values and initiating message passing. KEEpictures provides a quick way to build up a complex image by providing a set of basic pictures that can be used as building blocks for a complex graphic image. KEEpictures provides tools for customized mouse activated functionality. These interface tools are written with Common Windows, Intellicorp's Common Lisp solution for object-oriented graphics. After a particular application has been customized by the system developer, the user has the benefit of a rich graphic interface and mousable commands.

Advantages and disadvantages. KEE, as implemented in the 386 environment, provides an incredibly rich array of expert systems development tools. Applications that would have required a Lisp machine and extensive original code development just a few short years ago may now be completed with a less-expensive platform. The system now comes "bundled" with Interactive 386/ix (UNIX), Lucid Lisp 2.0.5, Interactive VP/ix (DOS), and Microsoft Windows 2.0. It is apparent that there is little that this system would not provide with

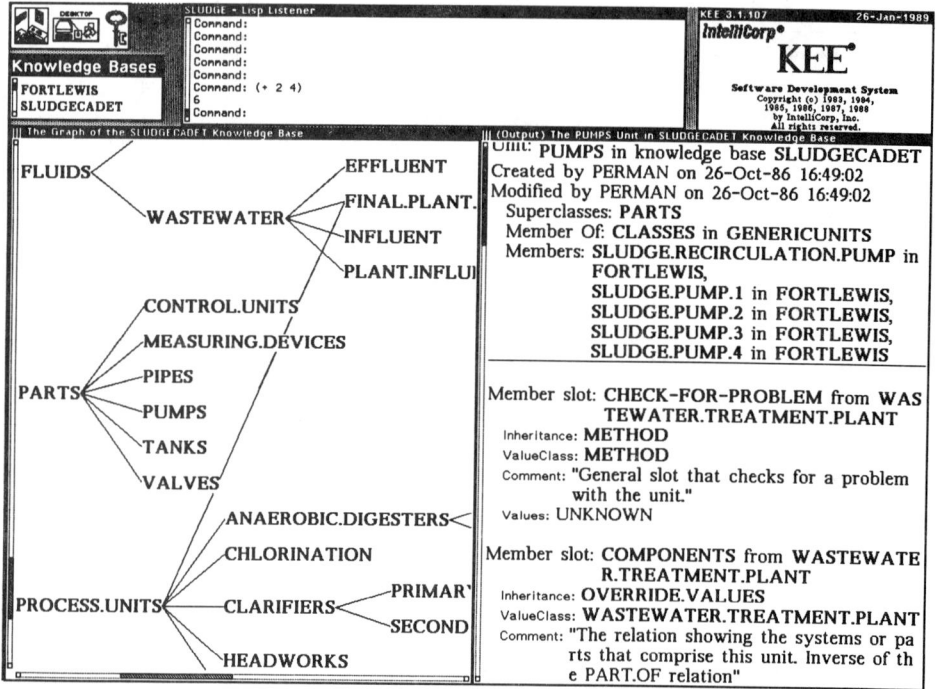

Figure 2.6. A second screen from the SLUDGECADET KEE application.
Notice the use of multiple windows being used for system development.
Included are a variety of tools on the *desktop,* including a *LISP Listener* win-
dow. Also note the use of slots, classes and inheritance.

respect to application development tools.

On the other hand, it is also clear that the system developer here is
unlikely to be the actual system user, except in specific cases. The system
developer in KEE is likely to be a trained knowledge engineer who is fluent in
Lisp programming techniques. It is also obvious that the total system cost,
including hardware, software and development time is likely to much higher on
this more advanced system.

Conclusions

We have examined a range of microcomputer-based tools for building expert
systems, ranging from small rule-based systems to more complex hybrid shells.
In making a choice about using one of these development tools for a specific
application, the knowledge engineer (or domain expert building their own

system) must use a variety of criteria for their selection. Waterman (1986) mentions criteria of power and sophistication, availability of support facilities, reliability, maintainability, and availability of required features suggested by the needs of the problem and application.

Some of these features were mentioned in this chapter in the sections on advantages and disadvantages of the different shells. For example, shells which allow only backward chaining should be used for problems where the number of goals is reasonably small and the possible outcomes are known. Forward chaining capabilities provide the more generalist approach to problem solving, where one first inquires in a general way about aspects of a problem. It is often just such a generalist approach that is required in planning and design problems. Shells that provide varied options for the chain of reasoning may be required in order to provide the basis for systems designed to solve planning and design problems.

Another specific criterion that planners may find of concern Includes whether graphic capability (both "pre-canned" and interactive) is necessary to convey certain information to the user. For example, systems for site selection or site planning might require such graphics for clarity. Another criterion particularly relevant to large-scale planning problems may be the requirements of accessing an external data base or performing complicated external computations. We have seen a variety of solutions to the database and computation problems in this review. These solutions include shells that provide these functions internally, and those which provide some link to an external application program or programming language. We have also noted that the nature and ease of the links to external programs varies between the shells. This criterion should be strongly weighted in the decision to use a particular shell if there will be frequent need for such external access. A final criterion might include the ease of system development, and the required presence (or absence) of a knowledge engineer in this development. Some of the tools described in this review are less suitable for the novice system developer without the aid of a trained Lisp programmer. Others are more suitable for quick prototyping and fast development of small systems.

In chapters 3 to 15 in this book, you will find many choices about system design that were made by the various developers of these early expert systems applications to urban planning. One of their most important choices was whether to write the system in a programming language, or to choose one of the available shells. As you read through this range of planning applications, you will be able to appreciate their application of these tools to a set of difficult problems.

References

Buchanan, B.G., and E.H. Shortliffe, eds., 1984. *Rule-Based Expert Systems: The MYCIN Experiments of the Stanford Heuristic Programming Project,* Addison-Wesley, Reading, MA.

Feigenbaum, E., P. McCorduck and H.P. Nii, 1988. *The Rise of the Expert Company,* Times Books, New York, New York.

Fikes, F., and T. Kehler, 1985. "The Role of Frame-Based Representation in Reasoning." *Communications of the ACM,* 28, no. 9, September, pp. 904-920.

Harmon, P., 1986. "Expert Systems-Building Tools." *Expert Systems Strategies,* Cutter Information Corp., 2 (8), 17-22.

Harmon, P., and D. King, 1985. *Expert Systems: Artificial Intelligence in Business,* New York: John Wiley and Sons.

Hu, D. 1987. *Programmer's Reference Guide to Expert Systems.* Indianapolis: W. Sams & Company.

Level Five Research, Inc., 1987. *Insight 2+,* Version 1.0, Indialantic, Florida.

Micro Data Base Systems, Inc., 1987. *GURU Reference Manual: Expert Systems for Information Management,* Micro Data Base Systems, Inc., P.O. Box 248, Lafayette, Indiana 47902.

Neuron Data Inc., 1987. *Nexpert Object Fundamentals,* Version 1.0, Palo Alto, California.

Stroustrup, B., 1988. "What is Object-Oriented Programming?" *IEEE Software,* 5, no. 3, May, pp. 10-20.

Texas Instruments, Inc., 1988. *Personal Consultant Plus: Reference Guide,* Texas Instruments, Inc. P.O. Box 2909, Austin, Texas 78769.

Waterman, A.D., 1986. *A Guide to Expert Systems,* Addison-Wesley Publishing Co., Reading, MA, 419 p.

PART TWO

Applications to Land Use and Transportation Planning

Introductory remarks by Lyna L. Wiggins

A recent book by Feigenbaum, McCorduck and Nii (1988) portrays an optimistic view of "The Rise of the Expert Company." Just a few years past the "dawn" of expert systems, these authors suggest rapid, and somewhat startling, increases in the use of these systems by innovative firms. These firms report a variety of gains due to their development of expert systems, including: (1) costs saved on internal processes; (2) speedup of problem solving; (3) preservation of the know-how of valued employees; (4) improvement of quality and consistency of decisions; (5) changes in the way the basic business works; (6) new revenues from new products; and (7) a stimulation of innovation. The authors of the chapters in Part Two of this book begin to address the possibility of "The Rise of the Expert Planning Agency." More particularly, they focus on the potential role of expert systems as decision-making aids for professional planners.

The role of expert systems in planning, relative to their use to date in private firms, will be more varied, subtle, and complex. It will also be more difficult to evaluate the relative usefulness and success of these systems. Feigenbaum, et al., report orders of magnitude gains in both efficiency of decision making and in the equivalent cost savings from these systems. They also suggest definite improvements in the quality of decisions made with the aid of these systems. The actual ability of the authors to measure quality improvement, however, is not clear. Constructing useful evaluations of these systems in private firms will prove a more complex task than is suggested in their optimistic presentation. The potential complexity of evaluating expert systems designed for planning is even greater.

One reason for the increased complexity of expert systems designed for planning is that most of them will attempt to solve problems of planning and design. Referring back to Chapter 1 of this book (Ortolano and Perman), we saw a possible classification of expert systems into 5 groups: (1) interpretation; (2) diagnosis and prescription; (3) design and planning; (4) monitoring and control; and (5) instruction. The large majority of the systems described by Feigenbaum, et al., as showing early success in these innovative companies, belong to the class of diagnosis and prescription. Many of the remaining systems developed by these companies are intended for instruction, interpretation, or monitoring and control. Very few of the systems attempt to solve the problems we think of as design and planning applications. There is a good reason for this. This is the most difficult class of problems to address with expert systems. This class of problems also requires the most complex set of programming capability to implement. (For example, most planning and design problems will require capability well beyond simple backward chaining.) The authors in this book are attempting to construct prototypes of systems for planning and design. They are necessarily complex because of the nature of the applications.

The four chapters in Part Two describe expert systems applied to problems in land use and transportation planning. Four themes emerge in these chapters. The first theme is that to be useful to planning practice the systems must be more integrative than the single-purpose systems used in many diagnosis and prescription applications. The second theme is that the complex nature of design and planning problems requires that expert systems be viewed quite clearly as decision making aids, not replacements for professional judgment. A third common point is that these aids to planning judgment must be heavily data-dependent, and must help the planner "get a handle" on useful information. A fourth theme, bringing the first three points together, is that to handle complex problems that are also heavily data-dependent the integration needs of these systems will require a combination of software components.

In Chapter 3, Wright describes a decision support system for land use planning. Specifically, the system is designed for a site selection problem. In the description of his prototype automated system, the need for an integrated solution is clear. Wright's approach is to combine components from geographic information systems (GIS), knowledge-based systems (KBS), and operations research (OR) into an integrated software prototype. To accomplish this, a combination of software components is required. For the geographic information system, an existing grid-based package is used. For the expert system component, he chooses to use a shell. The third component is an optimization model, implemented using a third software tool. After considering a number of possible choices of spatial optimization approaches, Wright selects one which allocates single uses to regions with the goals of minimizing cost and maximizing suitability. The expert system component of the prototype consists of rules which specify which land attributes for particular activities are important to meeting site selection criteria. The expert system acts as a "front-end" to the GIS, and the OR module receives input from the GIS. Together, these modules suggest an approach to an integrated solution of a complex planning problem.

In Chapter 4, Davis and Grant describe a system designed to aid local public-sector planners in developing zoning schemes. The final product is an "activity-allocation" table. The system (ADAPT) grew out of an earlier, more algorithmic, modeling effort, where the principal lesson was that an essentially political task like zoning requires that heuristic knowledge be directly incorporated into the system. The authors emphasize the theme that most of the relevant knowledge must come from the user, with the expert system providing relevant information, data about conflicts, and a record of the decisions made. Again we see the themes of integration (here between planner and system), and the need for large-scale data handling (here the system acts both as data base and recording device). Davis and Grant describe future extensions to the current system that allow for the incorporation of uncertainty, and the inclusion of input from multiple interest groups.

In Chapter 5, Brail describes a prototype transportation planning system. The system is a simplified version of the standard Quick Response System (QRS). Brail's emphasis in this chapter is on the use of one of the popular AI programming languages, Prolog, in the context of a planning application. His central theme is the necessity of data-intensive support systems for planning problems. Since a Prolog program can be viewed as an "intelligent" database, it may provide a useful tool for transportation planning, a substantive area where data entry and constant updating is central. Brail notes that the need for an expert system approach, rather than a standard solution, may be based on the need for the user to be able to ask hypothetical if-then questions more easily, and to allow the database to be changed more quickly within the system. Brail concludes that while Prolog may be a good tool for the needs of dynamic data base maintenance, it is less successful as a tool for completing the extensive computations required in transportation models. Again, we return to the theme that a solution which integrates various software tools is likely to be most successful.

In Chapter 6, Southworth describes the development of a real-time traffic monitoring and analysis system (RTMAS). This system addresses more of a monitoring problem, and less of a planning problem, than the other chapters in Part Two. But the common themes of data-intensity and the need for an integrated solution using a variety of software tools are striking in this application. The system being developed by Southworth is designed to record and monitor traffic flows, with the goals of classifying certain traffic patterns and warning of evacuations in times of emergency. In building into the system the ability to recognize particular patterns, the user (planner/engineer) must provide the information to the system about patterns due to unusual events (e.g., holidays, accidents, sporting events). The entire system relies on four major software components, including an expert system, an automated forecasting package, data acquisition software, and a program which classifies specific traffic patterns. The chapter ends with the description of actual case output from the system. Southworth reports that extensions to the system will include a predictive component to forecast traffic congestion "hot spots" before they occur.

References

Feigenbaum, E., P. McCorduck and H.P. Nii, 1988. *The Rise of the Expert Company*. Times Books, New York, New York.

CHAPTER 3

ISIS: Toward an Integrated Spatial Information System

Jeff R. Wright

Consider a region of interest to be partitioned into an arrangement of N non-overlapping geographical units numbered from $i = 1, 2, ..., N$. Every basic geographical unit can be characterized by a set of economic, environmental, and spatial attributes. Full and complete knowledge about the attributes for each unit is assumed. A subregion is any contiguous grouping of these units. The economic, environmental, and spatial characteristics of any subregion are a function of the collective attributes of the units comprising that subregion (Figure 3.1). Thus, for example, the total area of a subregion allocated for a land use is the sum of the individual areas of the unit comprising that subregion. The unit attributes may be represented as follows:

a_i = the area of unit i
c_i = the cost of acquiring and/or developing the land of unit i
s_i = the relative suitability of unit i for the proposed land use
d_{ij} = the straight-line distance between the centroids of units i and j

The cost of acquiring and developing unit i for a specific land use can be based on a number of market and non-market factors, including physical characteristics of the land, proximity to services etc. The suitability of a geographical unit, s_i, is defined to be a "measure of the physical capacity of a specific location (i.e. geographical unit) to support a specific land use" (Anderson, 1987). This determination is usually based upon natural characteristics of the land such as slope, soil type, and other physical factors. A number of mathematical models have been proposed to calculate suitability ratings as functions of these factors and underlying natural processes. The output from these models is a map of information depicting the relative capability of any given geographical unit to support a given land use. These models are discussed at length in Hopkins (1977a), Chapin and Kaiser (1979) and Anderson (1987).

The underlying assumption of this discussion is that the problem of selecting or allocating a particular set of geographical units as a subregion for a particular land use is appropriately partitioned into three sub-problems: 1)

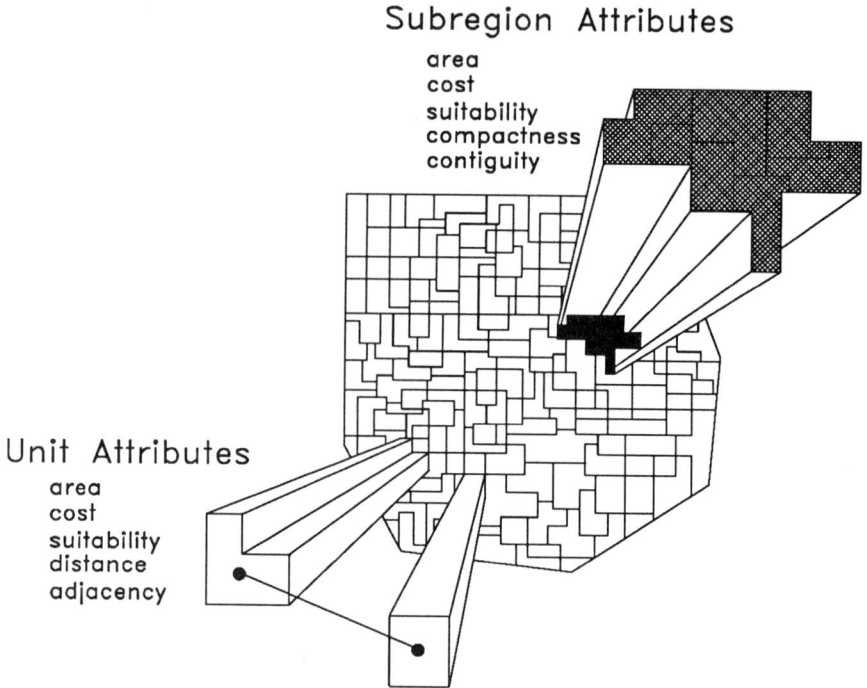

Figure 3.1. Unit and subregion attributes for the land allocation problem. Individual unit or cell attributes may be different from corresponding subregion attributes. Some unit attributes, such as cost, are additive when aggregated at the subregion level, while others, such as compactness, are not.

collecting, storing and managing data to be used in the determination of suitability for the intended land use, 2) the evaluation and analysis of data about each geographical unit to make that determination of suitability, and 3) the development of a strategy for selecting the set of units to form the subregion. Furthermore, three traditionally disparate technologies are most appropriate to address these sub-problems: 1) spatial data management using geographical information systems (GIS), 2) expert systems (ES) for determining land suitability, and 3) numerical optimization for identifying dominant land allocation strategies.

The Role of Geographical Information Systems

Recent years have seen a dramatic increase in the use of GIS for managing spatial data. While there exist several fundamentally different types of GIS technologies, the most straightforward representation is that offered by Figure 3.2.

Suppose we are interested in making a determination of the suitability of a particular region for a specific land use based on soil type. We might classify all relevant soil possibilities by assigning each a discrete category number (1, 2, ..., N). Of course, there exists a continuous distribution of soil type across any region. By overlaying a grid cell referencing system over our soil map, we are able to assign a corresponding discrete value through some averaging process. Because the location of each grid cell is known, an approximation of our original soils map may now be stored and manipulated as a rectangular matrix of category values. By performing the same approximation to attributes such as slope, vegetation, land use, ownership, etc., we can subsequently pose questions such as "How much land in the region having sandy clay soils and average slope of 5° or less is presently being used for commercial purposes?"

Land use planning and design, site selection, and other problems requiring the manipulation of geographic information have benefited from recent advances in geographic information systems (GIS). The incorporation of decision models into GIS has been discussed recently by several researchers, including Steinitz and Brown (1981), Tomlin and Tomlin (1982), Cederborg and Tosetti (1982), Gilbert et al. (1985), and Elmes and Harris (1986). The role of the GIS in the proposed implementation environment would be to store, generate, and manage all of the data required for the land use allocation model. This would include the computation of distances between geographical units, the determination of areas and adjacency relationships, and the storage and calculation of acquisition and development costs for each unit based on, for example, the slope and soil characteristics of the associated land. All of this information could be conveniently stored and/or calculated by a GIS and passed to the land use allocation model without manual intervention.

Because GIS are very efficient at the manipulation of geographical and spatial data, GIS systems are also ideally suited to support the task of assessing the suitability of land to support a particular land use.

The Role of Knowledge-based Systems

Most land use suitability assessment models are based on manipulations involving "factor maps". A factor is a unique geographical attribute such as slope of terrain, soil, or vegetative cover. Variations within a factor such as degrees of slope, soil classification or type or density of type of vegetation present are referred to as factor *types*. A factor map is a map showing the geographical location and distribution of factor types for a given factor. Factor maps can be combined in various ways to generate a composite map depicting, either graphically or numerically, the relative capability of any given geographical unit to support a specific land use. Hopkins (1977a), Chapin and Kaiser (1979), and Anderson (1987) discuss five major factor combination techniques. Four of these techniques are described briefly below. These techniques are: 1) linear

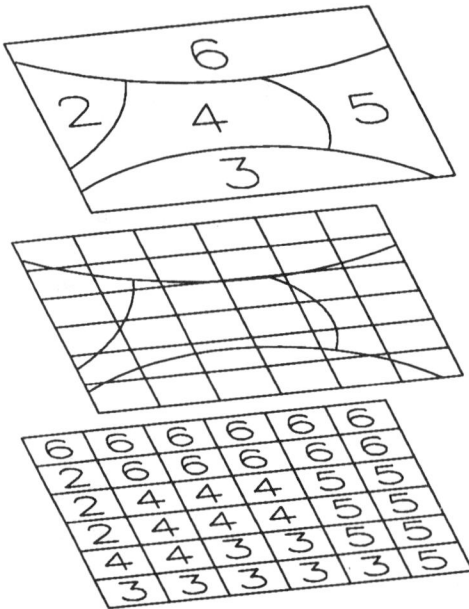

Spatial distribution of data exists as geometric shapes or polygons with a given attribute represented by a categorical assignment such as category 1 for soil type 1, category 2 for soil type 2, etc.

Using a grid cell partition of the region, an averaging of attribute data within each cell may be done.

The resulting matrix of categorical values may be stored and manipulated by a GIS program module.

Figure 3.2. Basic concept of raster GIS. Map layers can be produced using this procedure for a wide variety of physical and logical spatial attributes.

combination; 2) nonlinear combination; 3) direct assignment and 4) rules-of-combination.[†] The fifth technique, ordinal combination, has several major draw-backs as discussed in Hopkins (1977a) and will not be considered here.

1. Linear combination is a common factor combination technique. This method relies upon the decision maker to supply both a rating schedule and a weighting schedule. The rating schedule depicts the relative suita-bility of the various types within a factor. For example, slopes greater than 20% are rated 0, between 10% and 20% a 5, and less than 10% a 10. The weighting schedule depicts the relative importance of the various fac-tors in determining the suitability of a unit. For example, slope might be twice as important as depth to bedrock. The ratings for each type are mul-tiplied by a weight for each factor and the result is summed.

†   The terminology used is from Hopkins (1977a) with the exception of "direct assignment", a term used by Anderson (1987).

2. One of the drawbacks of the linear combination method is that it cannot handle interdependence between factors. The composite score for a unit might be conditioned upon the presence or absence of a factor or interactions between factors. If these interactions are well-defined, then it may be possible to define a nonlinear function that explicitly captures these relationships. For example, a soil loss rating can be easily computed from standard equations given information on land cover, slope and watershed shape. As such, nonlinear combination methods can be very useful when applied to specific components of an overall suitability assessment. However, in practice the relationships between factors are rarely well-understood and it is difficult to represent most interactions as mathematical functions (Anderson, 1987).

3. The direct assignment technique involves enumerating all possible combinations of factors, and assigning a composite score to each unique combination when all factors are considered simultaneously. The advantage of this method is that it requires the explicit consideration of interactions between factors. The disadvantage of this method is that the number of composite score assignments that must be made increases exponentially with the addition of new factors.

4. The rules of combination method involves the use of explicit rules to assign suitabilities to *sets* of types rather than to single combinations. Rules of combination are expressed in terms of verbal logic rather than in terms of numbers and arithmetic. For example, "if type 1 of factor X and type 2 of factor Y occur together in a region, then the rating of factor X type 1 overrides that of factor Y type 2." These rules are based upon an understanding of the underlying natural system being described rather than on a single set of relationships repeated for all combinations regardless of the specific types and factors being combined. If properly devised, rules of combination can handle interdependence between factors. Furthermore, they do not require an explicit mathematical relationship as do nonlinear combination methods.

    Suppose that an activity is to be assigned to a region and the suitability for any particular cell to support that activity has been judged to depend on slope, geology and forest density. Suppose further that complete categorical information for the region of interest exists in a GIS as discussed above and as presented as Figure 3.3. An example of a combinational rule that might be used to determine suitability might be:

```
IFMAP slope 1-3
ANDIFMAP geology 4-6
ANDNOTMAP density 4
THENMAPHYP 4 suitability
```

This rule contains the judgement that if, for any particular cell, the average slope falls into slope categories 1, 2 or 3, and the average geology is of type 4, 5 or 6, and the forest density is not of type 4, then a suitability of 4 may be assigned.

Additional rules may be employed to reflect additional judgements:

```
IFMAP slope 1-3
ANDIFMAP geology 4-6
ANDNOTMAP density 4
THENMAPHYP 4 suitability
!
IFMAP slope 6-7
THEN too steep
!
IFMAP geology 1-3 7-8
THEN unstable surface
!
IFMAP density 4
THEN severe forest impact likely
!
IF too steep
THENMAPHYP 1 suitability
!
IFNOT too steep
ANDIFNOT severe forest impact likely
ANDIF unstable surface
THENMAPHYP 2 suitability
!
IFNOT unstable surface
ANDIFNOT density 3-4
ANDIFMAP slope 3-6
THENMAPHYP 3 suitability
!
IF too steep
ANDIFNOT severe forest impact likely
ANDIFNOT unstable surface
THENMAPHYP 3 suitability
```

These rules may then be imposed across the spatial extent of the region of interest, and for all appropriate map layers, producing a new map layer reflecting spatial suitability for the target land use as shown in Figure 3.3.

Three general observations can be made regarding factor combination methods:

1. Factor combination is often best achieved through the use of an explicit set of rules and verbal logic (Hopkins, 1977a).

2. The selection of meaningful ratings, weights, and methods of combination is a complicated and tedious task to undertake manually. In short, decision makers are faced with at least three complex tasks: a) identifying the most relevant set of factors; b) deriving a meaningful set of ratings and weights; and c) identifying an appropriate functional form for combining factor maps that reflects the relationships between factors.

1	1	2	2	3
1	2	2	2	3
1	2	2	3	3
4	3	4	4	5
5	4	4	5	6

SLOPE

0 No available data on slope
1 Average slope 0°
2 Average slope between 1° and 2°
3 Average slope between 3° and 5°
4 Average slope between 6° and 9°
5 Average slope between 10° and 14°
6 Average slope between 15° and 20°
7 Average slope greater than 21°

0	0	0	5	5
5	5	5	5	5
5	7	7	7	7
7	7	7	7	3
3	7	7	7	3

GEOLOGY

0 No available data on geology
1 Average geology is metamorphic
2 Average geology is transition
3 Average geology is igneous rock
4 Average geology is sandstone
5 Average geology is limestone
6 Average geology is shale
7 Average geology is sandy shale

4	4	4	4	3
4	4	3	3	3
3	3	3	3	3
3	3	2	2	1
1	1	1	1	1

FOREST
DENSITY

0 No available data on forest density
1 Average density is non-forest
2 Average density is low density
3 Average density is medium density
4 Average density is high density

0	0	0	0	4
0	0	4	4	4
4	4	2	2	2
3	2	2	2	2
2	2	2	2	1

SUITABILITY

0 No available data on suitability
1 Average suitability is poor
2 Average suitability is fair
3 Average suitability is good
4 Average suitability is excellent

Figure 3.3. An example of rule-based inferencing across map layers. A suitability raster map layer is produced from empirical data contained in other map layers through a rule-based inferencing procedure.

3. Expert judgement is an essential ingredient of the assessment process. For example, the rating assigned to each type of every factor must reflect all of the characteristics of that type and the potential impact of the proposed land use or facility on those characteristics and vice-versa. In describing an assignment technique, Lyle and von Wodkte (1974) states: "Although this sequence provides an orderly way of defining and articulating judgements, it relies heavily on the knowledge of those formulating the model. Usually, it is necessary to consult with experts in various disciplines during this phase to render the judgements as reliable as possible."

The problem characteristics outlined above are indicative of a "good" problem domain for a rule-based expert system (RBS). A rule-based expert system is a computer program designed to use domain specific knowledge gathered from human experts to perform problem solving tasks in a manner similar to humans. Rule-based expert systems are separated into an explicit rule base describing general problem solving strategies, and a control program to search the rule base. This programming framework allows the RBS to apply domain specific knowledge to a problem and make *inferences* about problems in a more flexible manner than traditional algorithm-based programs. It also facilitates the implementation of explanation procedures that can be used to justify the rationale behind a specific problem solving approach. Furthermore, most RBS have ways of dealing with uncertainty, including fuzzy logic, Bayesian analysis, and a number of other ad hoc procedures.

There has been very little research on the role of expert systems technology in the field of land use planning. Han and Kim (1987) have reported on the development of a prototype expert system for site selection and analysis. However, the system does not use suitability assessment techniques nor is it designed to interface with a geographical information system. Only a few researchers to date have proposed integrating GIS and RBS. Among them, DeMers (1985) discusses the potential role of RBS in the classification of land according to agricultural economic viability, Smith and Pazner (1984) describe a knowledge based system designed to answer queries concerning geographic objects in large spatial databases, Jungert (1984) proposes a methodology for conducting efficient searches in spatial databases employing inference systems, and Coulson et al. (1987) discuss the role of *intelligent* geographic information systems in natural resource management.

The perceived role of an expert system in the implementation environment is that of an interactive *front end* to a GIS. The RBS would query land use planners for the specific characteristics and requirements of the land use that is to be allocated. Based on this interaction, the RBS would identify the relevant geographical factors and prompt the user for the ratings and weights of each factor and the degree of "certainty" associated with each. If the user was relatively confident about these values, then no further questioning on this factor would be required. However, if the user was not confident about the values, then the RBS would ask additional questions that might provide insight (either to the user or the RBS) as to the most reasonable set of ratings and weights. As the session progressed, the RBS would check for the consistency of ratings and weights based upon information available in the rule base and the user's previous responses. At the same time, the RBS would serve as a knowledge acquisition facility by which all information input by the user or generated by the RBS would be stored in the rule base for future use as necessary. At the conclusion of this interactive session, the RBS would recommend a set of ratings, weights, and factor combination strategy reflective of an expert understanding of the underlying natural system. These recommendations could be accepted, modified, or rejected by the user on an ongoing basis until the process converged on an acceptable set of values.

The Role of Spatial Optimization

The premise of land use allocation is that certain geographical regions are better suited for specific facilities and land uses than other geographical regions. This section reviews several representative mathematical models that are designed to allocate land uses optimally among specific land areas. In this research, these are referred to as land use allocation models, although they are sometimes referred to as land use plan design models in the regional science and urban planning literature. As distinguished from the widely-studied facility location models that treat optimal locations as points in a plane or along a network, land use allocation models treat optimal locations as two-dimensional areas.

For the purposes of this research, land use allocation models have been divided into two groups: those that assign multiple land uses optimally among a set of geographically *predetermined* regions; and those that allocate a single use to a single region, the boundaries of which are defined by the model. For convenience, models in the former category are referred to as multi-use (M-U) allocation models, and the latter as single-use (S-U) allocation models.

Multiple Land Use Allocation Models

M-U allocation models allocate multiple land uses among a set of predetermined, non-overlapping geographical regions. Region properties, such as area, shape, and location are fixed in these models. In general, the allocation of land uses to regions is governed by one or more utility objectives, and constrained by the supply and demand for land. Little or no attention is given to the continuity of land uses across adjacent regions. In most cases the objective of early M-U models was to allocate future land use growth in expanding communities to available land in order to minimize development costs. More recently, several authors have proposed the incorporation of multiple objectives such as transportation and accessibility considerations into these models.

One of the earliest land use allocation studies was conducted by Meier (1968) who discussed a linear programming model to plan expenditures for the acquisition of recreational land. This was a single objective model with continuous decision variables and included considerations of the total budget available for land acquisition, the demand for recreation, the supply of sites, the cost of acquisition, and the relative value of recreation at each of the available sites. The decision variables in this model were defined to be the area purchased for each recreational activity. The objective was to maximize the recreational "value" of the acquisition program subject to budget restrictions and land availability. Although no actual case study was conducted, the potential role of such a model in recreational land planning was emphasized.

In contrast to Meier's model that allocated land to a single use such as recreation, most M-U models allocate multiple land uses and are characterized by a high degree of interaction between those uses. For example, Ripper and Varaiya (1974) described a linear programming-based econometric model for the optimal allocation of multiple and interacting land uses in an expanding urban area. This model was based on an input-output model concerned with the import and export of goods between regions. The regions were defined by a rectangular grid of cells to which various land uses were assigned. The objective of their model was to minimize total transportation, production and import costs within the urban area subject to constraints on the production of commodities and labor, and the geometry of the regional transportation network. The model confirmed their hypothesis that a regional structure of concentric rings around a central city was an efficient land use pattern given the underlying import/export assumptions.

Another single objective, multiple land use model was proposed by Dyer et al. (1982). The authors used linear programming in a study of agricultural land planning to allocate twenty land uses (primarily agricultural) among predetermined regions. The objective of one model was to minimize the total land area allocated subject to constraints on agricultural and livestock production, urban land requirements, and land availability. A second and related model was constructed to assess the relative utility of allocating a particular land use to a particular region subject to the same constraints. The goal of both models was to evaluate current and future needs for the preservation of agricultural land in Canada. These LP models were the foundation of a land planning system known as the Land Evaluation Model (LEM) developed at the University of Guelph in Ontario.

Land use allocation models based on linear programming typically employ decision variables that prescribe the amount of land in each region that should be devoted to a particular use. If, on the other hand, this amount of land is determined in advance of the model formulation, then it is necessary only to decide whether or not a particular use is assigned to a region. This was the approach adopted by Correia and Madden (1985) in their formulation of a mixed integer programming model to identify alternative land use patterns in a Portuguese municipality. The constraints in this model pertained to minimum requirements on area for specific land use developments, maximum area availability, budget restrictions, and certain policy constraints, such as ensuring equitable expenditures between different regions. Two objective functions were proposed, although no attempt was made to optimize them simultaneously. One objective was to minimize the total budget necessary to attain certain land banking and other purchasing targets. The second objective was to maximize the area purchased subject to budget constraints. Considerable attention was given to the interpretation of the dual in the context of municipality expenditures and market prices.

In recognition of the fact that most land use allocation problems involve multiple and often conflicting objectives, several land use allocation studies have formulated models that address multiple objectives simultaneously. For example, in both studies of Bammi et al. (1976) and Bammi and Bammi (1979) several objectives were considered including : 1) minimize conflict between adjacent land uses; 2) minimize travel time; 3) minimize tax costs; 4) minimize adverse environmental impact; and 5) minimize costs of community facilities. The constraints in these models pertained to limitations on the amount of land available for each type of land use, as well as minimum acreage requirements for commercial and institutional uses. The models were solved using the weighting method of multiobjective programming (see Cohon, 1978) in which a single objective is formed as a nonnegative weighted sum of the individual objectives. By varying the sets of weights, alternative land use plans were generated for Du Page County, Illinois, one of which received the widespread endorsement of municipalities involved in the planning process.

Barber (1976) described a multiobjective land allocation model based on three objectives: 1) the minimization of land-development costs; 2) the maximization of residential accessibility; and 3) the minimization of transportation energy costs. Constraints on the supply and demand of land, and the ratio of various land uses in each district were included. The model was cast as a linear program and a noninferior solution was found through an interactive, iterative solution technique. The results of a hypothetical case study were presented.

Werczberger (1976) proposed the application of goal-programming to the problem of locating residential and industrial land uses taking air pollution into consideration. A hypothetical example was discussed in which the objectives were: 1) number of dwelling units; 2) air pollution standards; 3) fuel costs; 4) industrial development; 5) housing distribution; 6) construction costs; and 6) housing density. The author concluded with a discussion of the advantages and disadvantages of goal programming in environmental planning problems.

Linear programming is an exact solution procedure yielding allocations that are demonstrably superior to all others. One of the drawbacks of exact procedures is the requirement that the problem be stated in a relatively formal and mathematically rigorous format. Heuristic procedures allow for greater modeling flexibility but lack the ability to make precise statements about results. For example, Weil et al. (1975) described an improvement hueristic for locating multiple land uses among regions arranged in a regular grid configuration.

Given a starting allocation, the procedure examined neighboring solutions to see if any improvement in the objective functions was possible, and if it was, performed a simple swap. The procedure stopped when no further improvement was possible. The method was demonstrated on a planning problem in which 14 objectives were collapsed to a single weighted objective function.

Arad and Berechman (1978) described a multi-stage planning model designed to allocate a set of interacting land uses based on the balancing of costs and benefits. The primary focus of the model was on activity interactions between land uses and across regions. The objective function was designed to incorporate weights to allow planners to examine trade-offs between costs and benefits. The cost function represented the direct and indirect costs of locating a unit of activity in a given region. Similarly, the benefits were associated with locating a unit of residential activity in a given region. The model was solved using an improvement heuristic employing the Newton-penalty function method. An application of the model to a new community in New York was described.

Steinitz and Brown (1981) discussed the structure and function of a computer-based planning system composed of a database management component, a land use classification system, and a set of heuristic land use allocation models. The purpose of the system was to predict and evaluate future trends in land use, and to simulate the effect of alternative land use planning policies and options. A set of 7 land use allocation models, each focusing on a particularly economic sector - housing, industrial, commercial, public schools, conservation, recreation, and solid waste - were discussed. The housing model simulated the growth and spatial distribution of new residential development based on the objective of maximizing profit to developers and builders. The industrial model allocated industries to sites based on development costs and accessibility considerations. The commercial model located facilities on the basis of minimizing construction costs. The public schools model selected sites for new schools based upon accessibility, land cost, and site characteristics. The recreation model allocated recreation facilities according to site quality, cost, and the availability of funds. These and other models were used in growth management studies of Boston's South Shore region.

All of the studies reviewed thus far have been concerned with the location of land uses among a set of predefined regions. That is, the study region has been divided into subregions and then facilities/land uses are assigned to those subregions without regard to their specific location within those regions. Table 1 summarizes the major characteristics of the studies reviewed in this section. Although a number of other multiple land use allocation models have been presented in the literature, many have the same general features as those models reviewed in this section. In the next section, models that allocate a single land use to a region with no predefined boundary will be reviewed.

Table 3.1. Characteristics of Several M-U Allocation Models

Model Characteristic	1	2	3	4	5	6	7	8	9	10
multiobjective					•	•	•	•	•	•
heuristic								•	•	•
regular grid		•				•	•	•	•	
activity interaction		•				•		•		
environmental factors					•		•	•		
proximity factors					•	•		•	•	•
linear model	•	•	•		•		•			

1-Meier (1968)	6-Barber (1976)
2-Ripper and Varaiya (1974)	7-Werczberger (1976)
3-Dyer et al. (1982)	8-Weil et al. (1975)
4-Correia and Madden (1985)	9-Arad and Berechman (1978)
5-Bammi et al. (1976)	10-Steinitz and Brown (1981)
Bammi and Bammi (1979)	

Single Land Use Allocation Models

In contrast to allocating land uses among a set of predetermined regions, single land use allocation models construct the optimal subregion for a land use by a combinatorial "region-building" process. The area under consideration is first divided into an arrangement of basic geographical units. The subregion to which a single land use is allocated is defined by a contiguous grouping or aggregation of these units. In this way, S-U allocation models are similar to models of political redistricting and school rezoning in which a set of basic units are grouped into several nonoverlapping regions. The combinatorial nature of these models makes the identification of optimal groupings a difficult and computationally intensive task.

Wright et al. (1983) and Gilbert et al. (1985) have proposed multiobjective integer programming models for the problem of allocating a single land use to a single subregion by aggregating basic geographical units. For computational reasons, both studies assumed that the study region was divided into a regular grid in which all geographical units were uniform in size and shape (hence referred to as *cells*). Each study is discussed in detail below.

Wright et al. (1983) proposed a multiobjective integer programming model for the *land acquisition* problem. The problem was to select or acquire tracts of land for the development of a facility or land use. Because this was considered to be a multiobjective problem, the selection was governed by a number of factors, including the area, cost, shape (compactness) and economic value of acquired land. Several formulations were presented, but the emphasis was on a particular formulation based on the regular grid that provided the basis for a specialized and highly efficient solution algorithm.

Gilbert et al. (1985) discussed the development and implementation of another multiobjective discrete optimization model for the allocation of a single land use. The objectives considered in this allocation model were cost, proximity to desirable and undesirable land features, and the shape of the region allocated. The model and the method of solution were also based upon the assumption of a regular grid configuration where some cells were designated as infeasible for allocation, and others designated as amenity or detractor, depending upon whether it was associated with a desirable or undesirable land feature.

There are several drawbacks to both S-U allocation models. First, both models are based on a regular grid configuration. This assumption appears to be for computational convenience. For example, Wright et al. (1983) exploit the regular grid structure with the use of external border to predict the location of noninferior solutions. Furthermore, in the regular grid configuration all units are of the same area. Consequently, if a subregion of area K is to be allocated among N geographical units, complete enumeration requires the examination of "only" $\binom{N}{K}$ solutions. With the irregular grid configuration, the maximum number of solutions that could require examination would be 2^N. For a problem with K=5 and N=100, the former could require the enumeration of 10^8 solutions while the latter could require a total of 10^{30} solutions. Of course, the actual number of solutions that require examination is often only a small fraction of these amounts, but the difference between potential effort is considerable. The assumption of a regular grid is computationally advantageous but unrealistic, because real estate parcels and tracts of land are seldom uniform in size and shape.

Second, neither model addresses the issue of the impact of the facility on the land. The minimization of impact should be an important consideration in the siting of any major facility or land use. Third, the existence of multiple decision makers may preclude the use of multiobjective methods that normally rely on the existence of a consistent utility function. Although, Wright et al. (1983) employ a trade-off curve generation technique, it is not clear that their method is capable of handling moderately sized problems.

In the following section a model formulation is presented that explicitly addresses the concerns discussed above. The details of the model structure are presented in the following section including a short summary of computational experience with an explicit enumeration solution technique. For a more thorough discussion of this algorithm, the interested reader is referred to Diamond (1988).

An operational land use allocation model must be both well-defined and "solvable." A well-defined problem is characterized by: 1) a representation that permits explicit identification of possible alternatives; and 2) an evaluation function that permits an explicit preference ordering of alternatives. For a multiobjective programming model, a well-defined problem includes the following:

1. A well bounded definition of the physical and/or conceptual system being modeled.

2. A set of decision variables representing those features of the system that are controllable.

3. A set of constraints, expressed as functions of those decision variables, that represent limits on the system.

4. A set of operational objective functions that can be used to measure the performance of the system.

Each of these requirements is discussed in greater detail in this section. The issue of "solvability," though not independent of the formulation of a well-defined problem, is addressed in greater detail elsewhere (Diamond, 1988).

The statement by Wright et al. (1983) that "land ownership is inherently discrete and one is unlikely to sell a fraction of one's holdings simply because they display specific characteristics" characterizes the discrete nature of the S-U land allocation problem. In both the Wright et al. (1983) and Gilbert et al. (1985) studies as well as this research, binary (0-1) variables are employed to indicate the decision to include or exclude a basic geographical unit from a noninferior subregion. More formally, the decision to include or exclude a particular geographical unit can be represented by the variable x_i such that:

$$x_i = \begin{cases} 1 & \text{if unit i is included in a subregion} \\ 0 & \text{otherwise} \end{cases}$$

A subregion can be completely specified by the vector $x = \{x_1, x_2, \cdots, x_N\}$, consisting of the set of N 0-1 decision variables. In the following sections, the constraints and objectives of the problem will be modeled as an integer program using these variables.

Although the S-U allocation problem could be formulated with many different types of constraints, there are several "core" constraints that are particularly characteristic of the problem. These restrictions pertain to the contiguity, area, and shape of feasible subregions. These were primary considerations in both models of Wright et al. (1983) and Gilbert et al. (1985) and are also central to the formulation presented in this research.

Area. It is assumed that any proposed land use will have certain minimum land area requirements. Let $\underline{a}$ represent this minimum area of land. Given the parameters and decision variables defined in the previous section, the minimum area requirement can be mathematically expressed as:

$$\sum_{i=1}^{N} a_i x_i \geq \underline{a}$$

It is also assumed, for computational convenience, that an upper bound can be placed on the maximum area of land that would ever be required by the facility. If $\bar{a}$ represents this maximum area, then

$$\sum_{i=1}^{N} a_i x_i \leq \bar{a}$$

will restrict all subregions to be no greater than this area. This area upper bound constraint can be an effective means of reducing the amount of search required to find the set of noninferior subregions.

Contiguity. It is assumed that a feasible subregion must be contiguous. Contiguity implies that it is possible to travel from any point within the subregion to any other point within the subregion without leaving the subregion. Contiguity of a land use is important because it tends to minimize undesirable impacts of that land use upon surrounding areas and also reduces the costs of providing various services to the area. It is a central concept in political redistricting, school rezoning, and other region-building models.

It is difficult to model and enforce contiguity explicitly using conventional linear integer programming techniques. Wright et al. (1983) did not explicitly constrain solutions to be contiguous in their integer programming model. Instead, a trade-off curve for solutions guaranteed to be contiguous could only be generated for certain and specified values of external border. Gilbert et al. (1985) generated solutions without explicitly enforcing the contiguity restriction. Non-contiguous solutions were removed from further consideration only after they were generated and identified. Although contiguity might be considered a "complicating" constraint in the sense that it makes these models relatively more difficult to solve, it can be exploited to make the search for noninferior solutions more efficient (Diamond, 1988).

Compactness. Compactness is a relative term used to describe the shape of a figure. Measures of two-dimensional compactness can be found throughout the spatial modeling literature in which shape plays a key role. For example, compactness has often been used as a relative measure of political gerrymandering in political redistricting models (Garfinkel and Nemhauser, 1970). In the single land use allocation problem, it is desirable for an allocated land area to be relatively compact in shape because this will minimize undesirable impacts upon surrounding areas and will also reduce the costs of providing various services to the area. Wright et al. (1983) use a compactness function based upon the total perimeter of the acquired region. Gilbert et al. (1985) measure compactness as a function of both the perimeter and diameter of the region.

There is no universally agreed upon measure of compactness. Most measures have strengths and weaknesses. Some measures are unattractive because they are analytically cumbersome. Other measures are easy to compute but suffer from an inability to distinguish between two shapes that are clearly different (Wright et al., 1983).

Measures of shape are seldom separable or additive. When considering the interaction between adjacent geographical units, for example, the compactness of the individual units is of little or no use in determining the compactness of the region resulting from the aggregation of those units. Two highly noncompact regions could combine to form a compact region, whereas two compact units might result in a relatively noncompact region.

Despite the drawbacks of the various compactness measures and the lack of a standard criteria by which to measure compactness, it is useful to exercise some degree of control over the shape of the subregion allocated to the land use. In selecting a measure, it is desirable to choose one that is relatively independent of the total area of the subregion. That is, two figures with the same shape but different areas should yield the same compactness value. One relatively simple measure proposed by Horton (1932), D^2/A, is a function of both the area A and the maximum diameter D of a planar figure. The maximum diameter is defined to be the distance between the two most distant points in the figure. Using the current definition of decision variables, this compactness measure is given by:

$$\frac{max(d_{ij}x_ix_j)^2}{\sum_{i=1}^{N}a_ix_i}$$

where $max(d_{ij}x_ix_j) = D$ for all i and j included within the subregion, and $\sum_{i=1}^{N}a_ix_i = A$. By enforcing some upper bound, λ, on this ratio, it is possible to exercise some degree of control over the compactness of a subregion. The smaller this bound, the relatively more compact the figure. A circle, widely accepted as the "most" compact planar figure, provides a convenient lower bound of $\frac{4}{\pi}$.

$$\frac{max(d_{ij}x_ix_j)^2}{\sum_{i=1}^{N}a_ix_i} \leq \lambda$$

Formulation of Objectives

Objectives are operational statements of goals. An objective function is a surrogate for a goal that permits measurement of attaining that goal through a standard or criterion of performance. The allocation of a land use is inherently multiobjective in nature. The model in this study will be based on two objectives: 1) the minimization of costs for the acquisition and development of land in a subregion; and 2) the maximization of land suitability for the given use. These two objectives are consistent with those identified by Lindgren (1979) as being important in the siting of major facilities:

- *Minimize the costs of acquiring and developing the site.* The goal is to build and operate the facility (or land use) at the lowest possible cost. This includes the cost of acquiring and developing the land to support the facility.

- *Minimize the disruption to the natural environment.* This refers to the impact of the facility on the immediate and surrounding land and is directly related to the capability of the land to support that particular land use.

The first objective of minimizing the cost of acquiring and developing the site is relatively straightforward. For every piece of property that must be acquired and/or developed, a cost figure can be attached to that unit. The cost of acquiring the land might include the direct cost of purchasing the land from landowners, financing and administrative costs, lost property taxes, and relocation and demolition costs (Kitay, 1985). The cost of developing a unit of land can be based on physical characteristics of the land such as slope, depth to bedrock, soil type, vegetative cover, and other factors. In any case, once the relevant costs associated with each unit have been identified, the objective of minimizing total subregion costs can be expressed as a function of the summation of costs of individual units included in the subregion. Mathematically, this objective can be expressed as:

$$\text{Minimize } C = \sum_{i=1}^{N} c_i x_i$$

where c_i is defined to be the total acquisition and development cost associated with the geographical unit i.

The second objective, though not as well defined as the first, is often just as important in determining the success or failure of a land use allocation decision. Historically, failure to consider adequately the widespread and often irreversible impacts of certain uses on the land and environment has resulted in the abandonment of many facilities and the loss of millions of dollars (Williams and Massa, 1983). As discussed previously, the suitability of a particular piece of property for supporting a proposed land use, or what is the same thing, the potential impact of the facility on the land, can be established by various mathematical suitability models. For any particular unit of land i, a suitability index, s_i, can be defined such that for any two units i and j, $s_i < s_j$ implies that unit j is more suited for development of the proposed land use than is geographical unit i. The objective of maximizing environmental suitability can then be expressed as a function of these parameters. There remains the issue of defining this function.

Although there appears to be no formal research on this particular topic, it is clear that there are several ways to express overall subregion suitability as a function of individual geographical unit suitability. For the purposes of this research, the suitability of a subregion will be based on a *weakest link* principle;

that the suitability of any given subregion can be no greater than the least suitable unit within that subregion. This appears to be a more realistic function than a simple summation or weighted linear combination of suitability indices. In qualitative terms, a subregion comprised entirely of units that are only "moderately" suitable is probably more desirable than a subregion comprised of a number of relatively well suited units but also some very poorly suited land. Consider for example two potential sites for a landfill. Site A is predominantly flat and has good soil characteristics throughout except for a narrow band of highly permeable soil running across the width of the site. Site B is gently rolling terrain with a moderate soil erosion potential, but has no severe limitations otherwise. Using a linear combination method, it is possible that the former subregion (A) could be rated more suitable than the latter site (B). Using the weakest link definition, the moderately sensitive land (site B) would always be considered relatively more suitable than that containing environmentally sensitive areas.

The objective of maximizing suitability can be expressed as a function of the suitability of a given subregion. If the suitability of a subregion is defined as $\min_{\{i \mid x_i=1\}} \{s_i x_i\}$, then:

$$\text{Maximize } S = \min_{\{i \mid x_i=1\}} \{s_i x_i\}$$

represents the "maxi-min" objective of maximizing the minimum suitability of any given subregion for the proposed land use. Conversely, this objective can be interpreted as minimizing the maximum impact of the proposed facility on the land.

Although in theory the model and approach proposed in this study could be modified to accommodate any number of objectives, in an operational sense there would be computational problems with more than two or three objectives. For example, Cohon (1978) states that the number of possible solutions to a multiobjective programming model tends to grow exponentially with an increase in the number of objectives. This tractability problem is compounded by the discrete nature of the model—the number of possible solutions to most discrete optimization problems tends to grow exponentially with an increase in the number of decision variables.

Prototype Structure and Function

The Army Corp of Engineers is faced with the problem of scheduling various training activities on the land of installations around the country. Each training activity has specific demands with respect to the type of terrain, the amount of land area needed, and other resource requirements. Each training activity also has an impact on the land in terms of soil erosion, soil compaction, disturbance of vegetative cover, and other environmental factors. At each installation there are numerous possible sites that could support any given training activity. However, the Corps would like to schedule activities so as to minimize both the cost of conducting the activity on the land and the environmental impact of the activity on the land. A prototype integrated spatial information system (ISIS) is being developed to assist the Corp with the identification of sites appropriate for a given activity and this set of objectives.

The prototype ISIS will consist of the land use allocation model developed in this research, the geographical information system "GRASS" (Geographical Resource Analysis and Support System) developed by the Corps of Engineers (Westervelt et al., 1987), and "KES" (Knowledge Engineering System), a rule based expert system distributed by Software A & E (1986). The preliminary structure of the ISIS is illustrated in Figure 3.4.

As discussed previously, the role of the expert system is to interact with the user (that person responsible for scheduling training activities) and converge on an acceptable schedule of suitability ratings reflecting the potential environmental impact of a specific training activity on the land at a given site. The geographic information system provides the data storage and management capabilities as well as routines to graphically display information to the user. Information and instructions will be passed between both systems as well as the model in the form of communication files and system calls. The model will use the information from both the RBS and GIS to allocate land for the activity based on the minimization of cost and impact on land (maximization of suitability).

One method of linking the KBS to other software, as in the case of the ISIS prototype, is by *embedding* the KBS. Embedding means that all interaction with the knowledge base is done through programs that are written by the user. This is similar to having the KBS as a subroutine of another program.

The majority of the knowledge base for the ISIS system consists of a series of rules that specify which land attributes and land attribute categories are important in meeting the site selection objectives for a particular activity. These rules assign weights to each of the categories within each land attribute file for each of the site selection objectives. The weights provide a relative measure of each attribute category's influence on the objectives. The remaining rules in the knowledge base infer what size land area is needed for the land use based on the number of people who will participate in the activity.

The KBS is provided with the participant information as input to its inference engine. Following the rules in its knowledge base, the KBS determines how large the selected site should be and which land attributes and categories within each attribute are conducive to the site selection objectives. This inferred information is passed to a program that writes this information into a file. The format of the file is such that the data pertaining to the two objectives are kept separate.

The GIS module of the prototype is also designed to be transparent to the user. To accomplish this, the GIS is embedded in the prototype and accessed only through programs that interact with the other systems in the prototype.

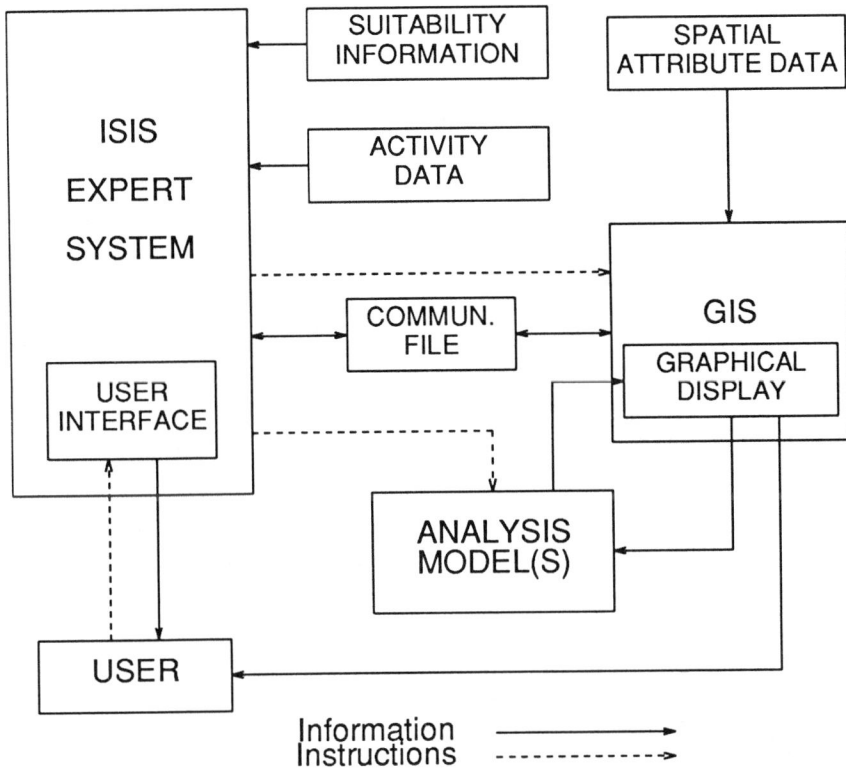

Figure 3.4. Schematic representation of the IGIS prototype. The executive program contains the expert system function for determining land suitability and for controlling the geographical information system and the land allocation optimization procedure.

In the ISIS prototype, a program that serves as a front end to the GIS reads the file that was written as the output of the KBS module. This program then queries the GIS database to provide the correct land attribute data files. This same program assigns the weights read from the input file to the categories of the appropriate land attribute. Cells in the data that exhibit a particular category of attribute have that category's weight assigned to the corresponding cell position in one of two arrays identical in size to that of the data files. Each array stores the weights pertaining to one of the objectives. Each subsequent attribute file queried and provided by the GIS in turn has its category weights added to the values in the proper positions in the proper array.

The data files stored in the grid-based GIS used in the prototype contain 457 rows and 397 columns. Each cell in the data file corresponds to a square land area that measures 100 meters on a side. To create data structures of a more practical size and number, the final function of the system's GIS module is to aggregate cells in the two objective arrays into larger *supercells*. Each supercell consists of a block of 16 by 16 cells representing an area of 2.56 million square meters or approximately one square mile. The weight assigned to the supercell is the mean of all the weights of the cells that comprise the supercell. The supercell weight value arrays are then passed as input to the OR module of the prototype.

The IP model used in the ISIS prototype is based on the multi-objective model presented by Wright et al. (1983). (Readers should see Wright et al. (1983) for complete description of IP formulation.) The parameters of the formulation are specified such that the selected site is composed only of contiguous supercells and the site is maximally compact. In that formulation, a zero/one variable exists for every supercell in the regular grid configuration. The model presents an optimal solution where the value of the zero/one variable will be one if the model designates that the corresponding supercell should be included in the site for the land use and zero if it should not be so included.

The input program to the OR module receives the supercell weight value arrays from the GIS module and uses them to make up the objective function used in the IP solution software system. The objective function for the prototype IP model is a combination of the two objectives presented in the problem scenario. These two objectives are combined into one objective function using the "weighting method" (Cohon, 1978). Each solution to the model represents an optimal solution for the weighted objective function combination used.

By repeatedly altering the weights used on the two objectives in the objective function and then resolving the IP model, a series of non-inferior solutions are obtained. A plot of these solutions with the value of one objective on the abscissa and the value of the other objective on the ordinate is called a *trade-off curve*. A trade-off curve is a graphical depiction of the trade-off between conflicting objectives.

In the ISIS prototype, a program alters the objective weights after each IP model solution and then calls on the IP solution software to solve the altered IP model. This process is repeated enough times to generate an approximation of the trade-off curve between the two objectives. The solutions from each iteration of this process are saved into a file to serve as input to a solution display program.

To use the ISIS prototype, the user must only enter the number of participants in the activity when prompted by the KBS embedding program. This program in turn provides this information to the KBS to begin the site selection procedure.

The next thing the user sees on the terminal screen is a display of the results of the site selection procedure. The results for each noninferior solution consist of a listing of supercell numbers that should be employed in the land use. The supercells are numbered beginning with the number one in the leftmost supercell in the top row. The numbering continues in unbroken sequence left to right across each row of the supercell matrix. The highest numbered cell resides in the lower right corner of the matrix.

The host computer used in the development of the prototype was a Gould PowerNode 9080. The 9080 is a supermini computer configured with a dual CPU (Central Processing Unit). The 9080 runs under UTX/32, Gould's port of the Berkeley 4.3 UNIX operating system.

The OR software tool used was the Zero-One Optimization Model (ZOOM) of the Experimental Mathematical Programming (XMP) LP package (Marsten, 1986). XMP is written in FORTRAN 77 and was originally developed at the University of Arizona.

The GIS selected for use in the prototype was the Geographical Resources Analysis Support System (GRASS) developed at the U.S. Army Corps of Engineers' Construction Engineering Research Laboratory (CERL). GRASS is a grid-based GIS first released in 1985 to assist land managers and training planners on military installations. GRASS is written in the C programming language and runs under the UNIX operating system (CERL, 1985).

References

Bammi, D. and D. Bammi, 1979. "Development of a Comprehensive Land Use Plan by Means of a Multiple Objective Mathematical Programming Model," *Interfaces*, Vol. 9, No. 2, pp. 50-63, Feb.

Cederborg, E.A. and R.J. Tosetti, 1982. "Site Selection for Low Level Radioactive Waste Disposal Sites," in *Waste Management '82: Proceedings of the Symposium on Waste Management at Tucson, Arizona*, ed. R.G. Post, Vol. 2, pp. 215-233, U.S. Department of Energy, Washington, D.C., March.

Chvatal, V., 1983. *Linear Programming*, W.H. Freeman and Co., New York, N.Y.

Clarke, K.C., 1986. "Advances in Geographic Information Systems," *Computers in Environmental Urban Systems*, Vol. 10, No. 3-4, pp. 175-184.

Cohon, J., 1978. *Multiobjective Programming and Planning*, Academic Press, New York, N.Y.

CERL (Construction Engineering Research Laboratory), 1985. *GRASS - Geographical Resources Analysis Support System*, U.S. Army Corps of Engineers, Champaign, IL.

Dangermond, J., C. Freedman, J. Harrison, and B. Bamberger, 1986. "The Conceptual System Design Process in Practice: The San Diego Regional Urban Information System Experience," in *Papers from the 1986 Annual Conference of the Urban and Regional Information Systems Association*, ed. R. Somers, Vol. 2, pp. 187- 201, Urban and Regional Information Systems Association, Washington, D.C., August.

Eilts, T., J. Wright, and M. Houck, 1984. "The Division Gunnery Model (DIGUM) as an Aid in Army Training Decisions", Report Number CE-HSE-84-3, School of Civil Engineering, Purdue University, West Lafayette, IN, July.

Gilbert, K.C., D.D. Holmes, and R.E. Rosenthal, 1985. "A Multiobjective Discrete Optimization Model for Land Allocation," *Management Science*, Vol. 31, No. 12, pp. 1509-1522, December.

Godfrey, K.A., 1986. "Expert Systems Enter the Marketplace," *Civil Engineering*, pp. 70-73, May.

Gooch, C.H., 1986. "Development of a Multi-Purpose Street Network to Support Mapping and Computer Aided Dispatching," in *Papers from the 1986 Annual Conference of the Urban and Regional Information Systems Association*, ed. R. Somers, Vol. 2, pp. 149-157, Urban and Regional Information Systems Association, Washington, D.C., August.

Harris, L.R., 1987. "The Marriage of AI and Data Base Technology," *AI Expert*, pp. 7-8, March.

Jensen, J.R., 1986. *Introductory Digital Image Processing*, Prentice-Hall, Englewood Cliffs, N.J.

Kubo, S., 1986. "The Development of Geographic Information Systems in Japan," in *Papers from the 1986 Annual Conference of the Urban and Regional Information Systems Associations*, ed. Rebecca Somers, Vol. 2, pp. 75-84, Urban and Regional Information Systems Association, Washington, D.C., August.

Marsten, R., 1986. *XMP/ZOOM*, Department of Management Information Systems, University of Arizona, Tuscon, Arizona.

Rehak, D.R. and H.C. Howard, 1985. "Interfacing Expert Systems with Design Databases in Integrated CAD Systems," *Computer-Aided Design*, December.

Tomlin, S.M. and C.D. Tomlin, 1985. "Computer-Assisted Spatial Allocation of Timber Harvesting Activity," in *Proceedings Auto-Carto 5*, American Society of Photogrammetry, pp. 677-686.

Tomlinson, R.F., 1984. "Geographic Information Systems-A New Frontier," in *Proceedings of the International Symposium on Spatial Data Handling*.

Waterman, D.A., 1986. *A Guide to Expert Systems*, Addison-Wesley Publishing Co., Inc., Reading, MA.

Wright, J., C. ReVelle, and J. Cohon, 1983. "A Multiobjective Integer Programming Model for the Land Acquisition Problem," *Regional Science and Urban Economics*, Vol. 13, pp. 31-53.

Yao, J.T.P. and K.S. Fu, 1985. "Civil Engineering Applications of Expert Systems," *Proceedings of the Fourth Ocean Mechanics and Arctic Engineering Symposium*, pp. 590-592.

CHAPTER 4

ADAPT: A Knowledge-Based Decision Support System for Producing Zoning Schemes[1]

J.R. Davis and I.W. Grant

In Australia, the authority to draw up and administer land-use plans is vested by state governments both in local governments and in special purpose state agencies. Broadly speaking, local governments control land-use change and development on privately owned land, and state agencies administer such publicly owned natural areas as national parks, crown land, and forests. Although public land planners can directly determine the type of land use and the extent of development in their jurisdictions, Australian local government planners have little ability to initiate desirable land uses.

Management plans for publicly owned natural areas generally consist of a strategic section, which allocates preferred land uses to all parts of the management area, and a more tactical section dealing with the ways by which the agency's resources other than land (personnel, equipment, funds) will be applied to bring about and maintain this land-use pattern (Compagnoni, 1984; Cocks et al., 1986). Local governments also draw up land-use allocation plans, but, since development initiatives lie with private landowners, these plans can only allocate land-use activities to contingent control categories. Typical control categories are "permit as of right," "permit only with the consent of the local authority," "permit only with the consent of specialist state government authorities," and "prohibit". A decision on each development proposal must await an examination of its features at the time it is submitted to the local authority (that is, at development control time). In spite of the similarities between the

---

1   The original version of this chapter was published in *Environment and Planning B: Planning and Design* 1987, 14:53-66.

strategic section of public land plans and private land plans, the following discussion is couched in terms of local government plans, since they form the genesis of the present work and constitute the majority of land-use plans in Australia. However, the ADAPT (A Decision Aid Planning Tool) computer program described in this paper can potentially be applied to planning of both private and public land.

Zoning schemes are the most common method of expressing local government plans. In a zoning scheme the local government authority states, in advance of development applications being submitted, the control category to be applied to each of the possible land-use activities in each part of the area being planned. The principal advantage of zoning schemes (rather than other methods of plan expression such as environmental standards) is their ability to balance the certainty required by developers considering the submission of development proposals with the flexibility of responses required by local government planners (Faludi, 1985).

It is useful, at this point, to describe a simple model of a land-use zoning scheme so as to introduce terms and procedures that are used in the ADAPT program. In this simple model, the region to be zoned is divided into a number of planning units, each relatively homogeneous in its use possibilities. Each planning unit is assigned a zoning which reflects the planning authority's land-use or development intentions for that area— typically, the same zoning will be assigned to a number of separate planning units. A zoning on a planning unit specifies a control category for each of the many activities that can occur on that unit. The appropriate control category to be exercised over a particular activity results from a consideration of broad community requirements (often encapsulated in state and regional land-use policies) and local concerns (local policies). Thus, in this simple model, the production of a local government zoning scheme can be formally stated to be the task of assigning each activity to a control category in each part of the planning region, in accordance with a set of predefined criteria.

Such a zoning scheme is generally presented as a map which shows the zoning to be applied at each location within the planning region, and an accompanying table which specifies the control category—activity combinations for each zoning (see Figure 4.1). In practice, the table is qualified with a set of clauses providing further details on each zoning; however, these clauses are ignored here.

Whereas in this model it is assumed that a zoning scheme is constructed by assigning individual activities to control categories, in many instances planners are constrained to use a set of predefined zonings. This typically happens when higher-level authorities, such as state governments, insist that certain zonings be used (for example, Laut, 1984), although it can also be a local requirement taken to ensure consistency across zoning schemes within one authority's jurisdiction (for example, Cocks et al., 1983). In these instances, the task of producing a zoning scheme can be reduced to that of assigning all parts of a planning region to one of these predefined zonings.

Zoning	Permit without reference to council	Permit with council consent	Prohibited
Rural 1(a)	agriculture; forestry	advertising structures; pig keeping; dwelling houses	shops; heavy industry
Rural 1(c)	agriculture	advertising structures; dwelling houses; rural residential allotments	heavy industry; pig keeping; shops
Industry		heavy industry; shops; dwelling houses	all other uses

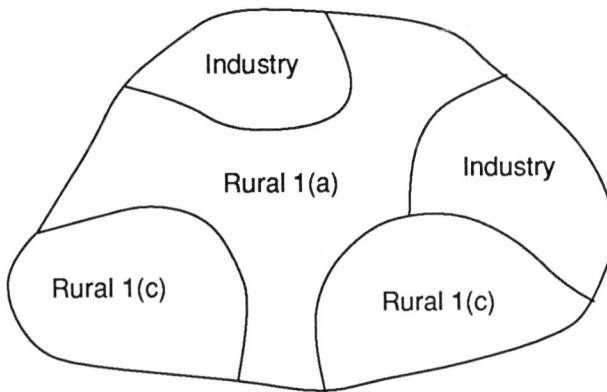

Figure 4.1. A typical local government zoning table and map. Each part of the local government area is assigned a zoning which describes the control that the local government wishes to exercise over each potential land-use activity. The zonings form the rows of the table and the control categories form the columns.

Decision-making Aides for Planners

There are potentially large, windfall financial gains (and losses) for landowners and developers as a result of decisions taken during the production of a zoning scheme. Consequently, the allocation of a control over each activity in each part of a planning region is not just a technically difficult task in multi-objective decision-making, but is also (in local government at least) a politically sensitive task.

The complexity arises from a number of causes, including partially stated criteria, incommensurate measures, incomplete knowledge, varying planning horizons, interpersonal trade-offs, and the balancing of personal and societal objectives. Given this milieu, it is not surprising that planners have tended to eschew computer technology in favor of their own judgment and experience.

Where computers have been used, they have generally been limited to data provision and less frequently to the application of models of natural processes (for example, flooding) and some human activities (for example, transportation modelling).

The major task faced by planners when drawing up any land-use plan is to assemble information from diverse sources (for example, special interest groups, political parties, scientific predictions, personal experience) and make decisions that best satisfy the multiple goals advocated by these sources. To do this, they must recognize and promote socially desirable opportunities and avoid socially costly hazards, assess and resolve as far as possible any inherent conflicts between the different parties, and be aware of the long-term and short-term impacts of decisions on the various parties using whatever data are available.

A number of algorithmic models of this decision-making process have been formulated (Timmermans, 1984). Perhaps the most sophisticated have used constrained multi-objective optimization techniques (Cohen, 1978); techniques which require considerable skill and computing power. Other widely used decision-making models employ a much simpler rating-and-weighting (RAW) technique (for example, Hill, 1968; Zetter, 1974; Anderson, 1981) wherein the ratings represent various criteria (biophysical and socioeconomic) for assessing the suitability of the alternatives, and the weightings express the relative preferences that society attaches to these criteria. When faced with competing alternatives, such RAW approaches rate each alternative against each criterion and then weight these ratings according to their perceived social importance. These weighted ratings are then summed to produce a numeric (cardinal) suitability score for each alternative; the alternative with the largest score is then judged to be "the best".

Generally, models that use the RAW technique are applied iteratively. The results from each application of the model are assessed and the criteria (and the means by which they are calculated) and the importances of these criteria are adjusted to obtain more acceptable solutions. RAW techniques were widely used during the early 1970s, especially for strategic planning (that is, structure plans) in the United Kingdom (for reviews, see Openshaw, 1977; Batey and Breheny, 1978; Barras and Broadbent, 1982), but they were seldom used for local government work. Openshaw (1977, page 27) comments that

> "At District level, where quantitative techniques have rarely
> been used because of computer hardware problems and the
> more intensely political nature of planning, the requirements
> are for a different kind of model."

The LUPLAN program (Ive and Cocks, 1983), developed in Australia after a major regional planning exercise on the South Coast of New South Wales (Austin and Cocks, 1978), uses RAW to allocate either the "best" land use or the "best" zoning to each part of a planning region. There have been four local government applications of the program (McDonald and Brown, 1981; Compagnoni et al., 1982; Laut, 1984; Davis and Ive, 1985), three of which involved evaluations of the suitability of the program for the production of local government zoning schemes. Compagnoni (1986) provides a more general critique of the use of LUPLAN for land-use planning. Since the ADAPT program was developed as a result of the second of these applications, it is useful to summarize some of the practical difficulties encountered during this exercise (Davis, 1985b); difficulties that influenced the design of ADAPT.

A number of the difficulties resulted directly from the use of the RAW model of decision-making; a model felt by the politicians, planners, and researchers involved in the exercise to be an inadequate representation of the realities of drawing up zoning schemes. There were two general areas of concern. First, the unidimensional weightings were a very simplistic method of representing what were, in reality, very complex judgments of social preferences involving varying time horizons, multiple interest groups, and uncertain effects. In addition, LUPLAN required that the weights be set prior to each iteration of the program, that is, that constant marginal values be used for these measures of importance. Although the cyclical application of the program does allow the planner to explore the effects of marginal changes to the set of weights, this was too indirect and too cumbersome a process to be useful in practice [a recent addition to the program allows the planner to explore the consequences of altering the weights on selected mapping units (Ive and Cocks, 1985)]. Second, the simple addition of the weighted ratings used in the RAW model to provide a numerical measure of the "best" alternative was unrealistic. In reality, the various criteria were often highly nonlinear and interdependent.

In the past, the use of arithmetic operations (even if they were nonlinear and constrained) to model human decision-making gave rise to practical problems. For example, the precision implied by calculating cardinal values for land-use suitabilities did not reflect the often high degree of uncertainty attached to the data, the methods of calculating the ratings, and the weightings.

The other major criticism encountered related to the relative roles of the planner and the program. Because of the responsibility borne by the planner, it was important that he or she remained the decision-maker and that the program acted as an assistant. For example, the planner needs to be able to receive explanations for conclusions advanced by the program.

However, in common with many algorithmic decision-making programs, it is difficult to obtain understandable explanations from LUPLAN because the issues being considered have been reduced to cryptic numbers and symbols. This "black box" nature of LUPLAN was also seen by McDonald and Brown (1981) as a deterrent to its use by local government planners. However, these shortcomings neither imply that LUPLAN is unsuited for other planning tasks nor that computer-assisted decision-making is without benefit in the production of zoning schemes. For example, in the above exercise, the program requirement that planners explicitly state their criteria for evaluation, its potential for streamlining data collection, and its ability to assess the success of a zoning scheme against the allocative criteria were all commented on favorably by the clients. Rather, the shortcomings arise from "categorizing the ... problem in solution terms [that is, RAW] before the problem has been sufficiently studied and understood" (Compagnoni, 1986, page 342).

Because the source of these difficulties was so deeply embedded in the design of LUPLAN, it was not possible simply to modify the program for local government planning. Instead a new program, ADAPT, was specifically devised to aid planners with this work.

The difficulties with applying a simple algorithmic approach to the production of a zoning scheme mirror more general concerns expressed during the 1970s with the design of management-information systems. The programs developed in response to these concerns—decision support systems (DSSs)—possess many features in common with ADAPT. For example, DSSs draw upon techniques from artificial intelligence research (principally from knowledge-based systems), and these techniques are similar to some of the more heuristically designed features of ADAPT. In the next two sections the relevant features of DSSs and knowledge-based systems are briefly summarized. This discussion is followed by a detailed presentation of ADAPT.

Decision Support Systems

Over the last twenty years the use of computers in business applications has expanded from simple routine transaction recording, to reporting and control systems, to integrated management-information systems that possess some programmed decision-making capability. Automating this process turned out to be more difficult than anticipated, largely because the development of management-information systems had been technology driven with insufficient attention being given to the modus operandi of organizations and the roles of individuals within them. As Simon (1981, page xii) says,

> "Many important decisions have qualitative components that do not lend themselves easily to the calculus of real numbers... Much decision-making leads to complexities beyond the reach of optimizing techniques, requiring approaches that are more heuristic than algorithmic."

After this realization, the focus of research into management-information systems shifted from the information base and the means of accessing it, to the corporate decision-making process itself and the role of the individual decision-maker (Klein and Hirschheim, 1985). Apart from the development of a different class of program that reflects this change in emphasis—DSSs—the first steps have also been taken towards constructing a theory of institutional decision-making on which the development of DSSs can be based. This theory, however, is not discussed in this paper: a full description is provided by Bonczek et al. (1981).

Although there is no widely accepted definition of a DSS, there is general agreement on the main characteristics that it should possess. First, it should be deliberately designed to deal with the unstructured tasks that comprise management, particularly at senior levels. Simon (1960, page 6) describes unstructured (or "unprogrammed" in his terminology) problems as being those where there is not a "cut and dried method for handling the problem, because it hasn't arisen before, or because its precise nature and structure are elusive or complex, or because it is so important that it deserves custom-tailored attention".

A second characteristic relates to the use of models in decision-making. Bonczek et al. (1981) describe an increase in the structure of decision problems as one moves from the top to the bottom of an organization and a correspondingly increasing opportunity to apply algorithmic models of these decisions. Since structure is a prerequisite to the development of algorithmic models, it is futile to expect to be able to apply this type of model to unstructured problems. Instead, Simon suggests that it may be possible to apply heuristic rules of procedure and general problem-solving strategies to them.

Third, and related to the above, is a concern to construct programs that support decision-makers with whatever information is relevant, rather than attempt to replace their capabilities with a model of the decision-making task. At present, this can only be accomplished on a system-by-system basis. But as the theory of DSSs develops, it may be possible to draw some general conclusions about the design of such systems.

Fourth, DSSs should support human-initiated and machine-initiated queries, particularly ad hoc human-initiated ones. These queries should be couched in language that is natural to the problem, to improve the effectiveness of the interactions. Klein and Hirscheim (1985) suggest that a more fundamental reason for requiring a natural language interface is to promote the transmission of the "personal knowledge" (Polanyi, 1962) on which managers must ultimately rely when making decisions about unstructured problems.

DSS researchers have identified artificial intelligence as being the source of techniques needed to operationalize many of the above features (linguistics and database management are two other research areas). In particular, knowledge-based systems have provided the means to obtain, store, and apply the heuristic knowledge required in unstructured decision situations and the natural language systems necessary to provide the human-machine interfaces required.

Knowledge Systems

There has been interest in artificial intelligence (AI) since the earliest days of computing, but only recently has this branch of computer science given rise to such practical products as industrial robots, speech synthesizers, and expert systems. The last are computer programs that, from a user's point of view, closely emulate the major facets of expert human performance in a particular problem area. [See Barr and Feigenbaum (1981), Winston (1984), and Charniak and McDermott (1985) for a general introduction to AI and Michie (1982), Hayes-Roth et al. (1983), and Weiss and Kulikowski (1984) for discussions on expert systems.]

However, expert systems are simply the most widely publicized members of a broader class of AI programs—knowledge-based systems—which attempt to emulate the type of human performance that is based upon acquired knowledge (not necessarily expert knowledge). Thus, a computer program such as HEARSAY-II (Lesser and Ermin, 1977)—a speech-understanding program—is an example of a knowledge-based system that can emulate a nonexpert human function. However, expert systems remain the most widely studied and applied class of knowledge-based systems, and much of the following discussion has resulted from the experience gained with them.

Because these systems have only recently been applied to practical tasks, there has still not emerged either a comprehensive classification of problem types or a general statement on how to design knowledge-based systems for these different problem types. Consequently, knowledge-based systems tend to be separately tailored for each application. However, in recent years there has emerged some consensus on the characteristic features of such systems (Brachman et al., 1983) and those features that are relevant to the ADAPT program are now discussed.

Separation of specialized knowledge. Knowledge-based systems maintain a clear separation between the specific knowledge that they possess about a problem domain and the mechanism for applying that knowledge to an instance of the problem. In this, they can be contrasted with traditional procedural computer programs which intermingle knowledge about the problem domain (often in the form of equations such as $y = ax + b$) with control information for applying that knowledge (for example, procedural statements in FORTRAN, such as GOTO, WRITE).

Production rules (Davis and King, 1977) which are simple condition-conclusion or condition-action statements have emerged as the most widely used method of representing domain knowledge, largely because of their modularity and ease of understanding (Barr and Feigenbaum, 1981).

Ease of modification is one advantage of maintaining the knowledgebase separately from the application mechanism. In general, modifiability is important during the construction and verification of the knowledge base, and in nonexpert knowledge-based systems (where the user may possess reliable and accepted expertise) it is likely to become important during the use of the system.

The simple syntax originally proposed for production rules by Post (1943) limits the complexity of the knowledge that can be represented and so a number of richer syntaxes have been devised in an attempt to increase the usefulness of this representation. The syntax used in production rules in ADAPT will be specified later in this paper.

Use of near-English. Natural language processing (both understanding and generation) is one of the topics being actively researched by computer scientists (for example, Charniak and McDermott, 1985), and designers of knowledge-based systems have used some of these results in their attempts to build systems that behave "naturally". Natural language generation is seldom used within knowledge-based systems for explanation. However, parsers are widely used to allow queries to be carried out in near-English.

Explanation and justification. It is commonly asserted that the ability to explain their methods of reasoning (self-knowledge) is a central feature of experts and consequently of expert systems (for example, Brachman et al., 1983). In general, the need for explanation in a knowledge-based system is determined by a number of factors, amongst which are the user's expertise relative to that of the system, and the degree to which the user is required to defend decisions that are (at least partly) based upon the conclusions drawn by the systems. In the present case, it is asserted that the planner would feel competent to query decisions that are based on knowledge from the system. In addition, planners' decisions in drawing up a zoning scheme possess significant financial implications and they must be able to justify these decisions to those affected.

Calls for explanation can generally be made both during a consultation and after the system reaches a result. Thus the user can request information from the system either to condition a reply to some question from the system, or to help review the line of reasoning followed by the system.

Uncertainty and incompleteness. Because the elements of knowledge brought to bear on a problem are generally of varying reliability, most knowledge-based systems allow the builder to incorporate some measure of reliability with each element when the knowledge base is constructed. During a consultation, the application mechanism uses these measures to assign a degree of reliability to any inferred decisions. Quite commonly, some essential knowledge needed for decision-making will be missing from the knowledge base and, when this occurs, the system will inform the user that it is unable to reach a decision. In this, knowledge-based systems differ from most traditional procedural programs which are based upon models assumed to have wide domains of applicability; that is, procedural programs will invariably provide an answer as long as the user supplies values for the parameters in the model.

The ADAPT Program

The ADAPT program (originally described by Davis, 1985a) is structured around three basic premises. First, although it is possible to automate decision-making in some relatively simple situations, it is necessary to rely upon the planner to resolve more complex problems that arise during the production of a zoning scheme. In particular, when there is either insufficient knowledge recorded or when the knowledge implies conflicting decisions, the planner is better placed to make decisions than is the program. Second, the system should provide information to assist the planner to resolve these more complex problems. Third, it is important that the reasons for decisions taken by the program or by the planner be recorded, not only so that the planner can review and improve emergent plans, but also so that development control planners can assess the suitability of development applications against the premises that underlie the zoning scheme. In short, ADAPT assigns control categories to activities in cases where it has the necessary knowledge and there are no competing alternatives. ADAPT also identifies cases where there are either competing alternatives or missing knowledge in order to provide the planner with relevant information on these alternatives. Finally, ADAPT keeps a record of decisions taken and their underlying reasons. The program operates in four consecutive stages culminating in an "activity-allocation" table which the planner can finalize into a zoning scheme in a fifth, manual stage.

Stage 1 — Knowledge Acquisition

In the first stage, the planner collects policies that are relevant to the production of a zoning scheme. In many Australian states, state and regional authorities issue such policies, although they are generally rather broadly worded. However, the richest sources are the local government authority itself and local interest groups. Such policies may not be formally worded or even committed to paper, but they form the basis for much of the decision-making at local level, including the production of zoning schemes.

ADAPT requires that these policies be expressed in a set of formal statements (rules) that relate the elements in the previously described model of a zoning scheme:

```
Assign   or   exclude   (activity)   to   (control
category)  on  all  (planning  units)  where  (cri-
terion) is true.
```

Thus typical rules would be:

```
Assign  rural  residential  allotments  to  "permit
with council consent" if agricultural capability
is class 3 or 4.
```

```
Assign pig keeping to "as of right" category if
current zoning is rural 1(a).
```

```
Exclude  pig  keeping  from  the  "as  of  right"
category if distance to town is less than 5 km
and effluent disposal capability is low.
```

At present, the syntax of the rules is rather restrictive with the criterion being constrained to a set of simple triplets (variable, operator, value) linked by the logical connectives OR and AND. The operator in each triplet must be one of "equal," "not equal," "greater than," or "less than," and the value must be a whole number. The conclusion of each rule must be the assignment (or the exclusion) of a single activity to a single control category.

The source of the rule (such as the state policy it was derived from) and descriptive text about the rule can also be attached. Note that the knowledge expressed in these rules is normally available to the planner, and so ADAPT is properly classed as a knowledge-based system rather than an expert system.

The set of rules need to be neither complete nor consistent. As each rule is entered, ADAPT parses the rule and draws up cumulative lists of the activities, the control categories, and the variables (and the values each variable can take). The planner can display and modify these lists during any stage through an editor which supports SHOW, ADD, DELETE, and CHANGE commands. If desired, he or she can commence building these lists prior to entering the rules.

The rules are recorded separately to the ADAPT program, and can also be modified with ADD, DELETE, UNDELETE, and CHANGE commands. During the first stage, the planner also supplies ADAPT with a list of planning units with each unit being identified by a numeric code. Again this list can be displayed and modified during this and subsequent stages of the program.

Stage 2 — Data Collection

The second stage is one of data collection and entry. The planner uses the list of data variable names, constructed in the previous stage, to identify those data items needed to apply the rules. Data collection is time consuming and this purposive approach makes this task as efficient as possible. It is not necessary that values be obtained for all data items on all planning units, as ADAPT will operate with incomplete data sets.

The data values can be entered into the database in any order because ADAPT keeps track of missing values. Thus, data can be entered by variable, by planning unit, or arbitrarily. Again, the SHOW, ADD, DELETE, and CHANGE commands allow the user to inspect and alter the data values at any stage of the program; that is, the data values are treated as another list.

Stage 3 — Assignments

The third stage is essentially a matching exercise. The program inspects each planning unit sequentially and, if the values of all the variables in the criterion of a rule match the values of those variables on that planning unit, then the activity is assigned to the control category specified in that rule on that planning unit. Thus, for the first example above, rural residential allotments will be assigned to the control category "permit with council consent" on all planning units where agricultural capability is recorded as being class 3 or 4. The program records the rule(s) that caused the activity to be assigned to the control category, together with any textual information that is attached to those rules. Assignments can be complicated by duplications, conflicts, and omissions; we will discuss these in turn.

Occasionally, two or more rules may allocate an activity to the same control category. ADAPT assigns the activity and records the multiple rules that caused the assignment.

When two or more rules attempt to allocate the same activity to different control categories, the program detects the conflict and requests the planner to resolve it and to supply a reason with the decision. The planner does not necessarily have to choose one of the categories suggested in the rules. To help resolve the conflict, the planner can request further information from ADAPT through both the SHOW command and arrange of WHAT and WHERE questions. WHAT questions refer to assignments already made on the current or earlier planning units. For example: What category was small rural allotments assigned to on planning unit 5? WHERE questions establish the planning units on which criteria are true. The criteria can be either the values of variables recorded in stage 2 or assignments; WHERE is it true that slopes are greater than 30% and small rural allotments have been assigned to the prohibited control category?

If, even after acquiring this supporting evidence, the planner is unable to resolve the conflict, ADAPT will note that this decision has been deferred and move onto other decisions. If the "assign" rules unambiguously propose a category for the activity, then the "exclude" rules are checked for any conflict with the proposed assignment. If there is conflict [in the previous example rules, when a planning unit is presently zoned rural 1(a), is less than 5 km from town, and has low effluent disposal capability], ADAPT again requests the planner to resolve it. If there are only "exclude" rules applying to a particular activity, then ADAPT will not assign the activity to a category (even if there is only one category unexcluded). The user is always requested to resolve these situations.

More commonly, there is no rule available to determine the control to be exercised over an activity on a particular planning unit. Before stage 3 commences, the planner has the option of providing the program with a default control category for each of the activities and, when no rule is available, ADAPT assigns the activity to the default. Thus, the common practice of including the phrase "all other activities" in the "prohibited" control category of a zoning scheme can be mimicked by establishing this category as the default for all activities not explicitly assigned to a control category by the rules. If a default has not been set for a particular activity, ADAPT will ask the planner to decide on the appropriate control category. Again the planner can either request further information or defer making a decision.

ADAPT records the control category that each activity has been assigned to, together with the rules (both "assign" and "exclude") that drove the assignment and the source of the assignment (that is, rules, default, or planner). If the planner's judgment was the source of the assignment, then any reasons advanced for the decision are also recorded.

Stage 4 — Review

In the final stage of the program, ADAPT displays the allocations that have occurred in an "activity allocation" table of the form shown in Figure 4.2. Each row of the table displays the controls to be exercised over development applications for each land-use activity on a single planning unit, that is, a separate (not necessarily unique) zoning has been developed for each planning unit. From this table, the planner can draw up a map which shows the spatial distribution of these zonings.

ADAPT has been designed to allow the planner to review this table and map and examine unsatisfactory assignments. Shortcomings resulting from the assumption that a plan covering a local government area can be produced from decisions made on a number of separate planning units will be manifest as undesirable cumulative effects (for example, not enough land being available for small rural allotments) and undesirable spatial arrangements (for example, industry being a "permitted" activity in a planning unit adjacent to small rural allotments). Additionally, such a review is likely to uncover some missing or incorrectly worded rules that gave rise to undesirable assignments.

The program memory of the reasons behind the assignments provides the planner with necessary information during this review stage. Thus, he or she can discover whether a particular assignment resulted from the operation of one of the rules, from a default control category being assigned, or from a decision by the planner (and the reasons underlying that decision). Additionally, the planner can ask the questions about data values and assignments described in stage 3.

Lastly, the planner can inspect the degree to which the rules (and hence their parental policies) have been used. This is provided through a simple count of the number of times each rule was used during the assignment stage. Thus, if it is important that the intentions of certain policies be incorporated in the plan (perhaps because they have been legislated by a higher level authority), then the planner can readily assess the extent to which the corresponding rules have been used.

The descriptive text attached to each rule in stage 1 can be examined to establish the significance of these rules. If, as a result of this review, the planner wishes to reformulate the activity-allocation table, he or she can (after modifying the rule set, the lists of activities, controls, data variables, data values, and planning units) rerun the program for some or all of the planning units.

Planning unit	Permit without reference to council	Permit with council consent	Prohibited
1	agriculture; forestry	advertising structures; dwelling houses; farm buildings	industry; residential flats
2	agriculture		advertising structures; dwelling houses; farm buildings; forestry; residential flats; industry
3	agriculture	farm buildings	advertising structures dwelling houses; forestry industry; residential flats
4	agriculture; forestry	advertising structures; dwelling houses; farm buildings	

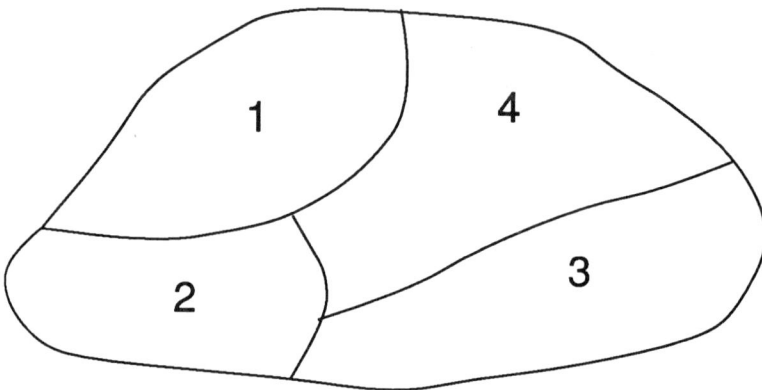

Figure 4.2. A simplified activity-allocation table. Its structure is similar to the typical zoning scheme illustrated in Figure 1 except that the rows consist of planning units rather than zonings. By amalgamating identical rows (for example, rows 1 and 4) and, if desired, similar rows (for example, rows 2 and 3) this table can be transformed into a zoning table.

Stage 5 — Zoning Scheme

In the final stage, the activity-allocation table and map (after the planner has completed any necessary revisions) are transformed into a zoning scheme by the elimination of duplicate (and possibly near-duplicate) rows of the table. This task is left to the judgment of the planner, again because of the complexity of the decision-making involved, although the planner can request the program to supply information to help with this simplification. For example, he or she could inspect the reasons that lay behind the assignments in two near-identical rows to decide whether it is reasonable to combine them. Thus, in Figure 2, rows (that is, planning units) 1 and 4 would be amalgamated and rows 2 and 3 could be amalgamated if they were judged to be similar enough.

In the present version of ADAPT, the reasons that underlie these amalgamations are not recorded, but it would be consistent with the philosophy of the program at least to record these reasons, and possibly provide the planner with rules that may assist with the amalgamations.

The concatenated table and the corresponding map now constitute a zoning scheme that has been built up from decisions taken (both by computer and by planner) about the appropriate controls to be exercised over individual activities in each part of a planning region.

Further Work

Two characteristics of the problem domain are not addressed in the present version of ADAPT. First, it does not place a degree of reliability on decisions taken or on information supplied; that is, it treats all data, recorded knowledge, and input from the planner as totally reliable. Second, ADAPT has no facility to examine the degree to which the different interest groups have been recognized in a zoning scheme.

Both these shortcomings are being rectified in a version of the program currently in development: the first through the use of techniques from expert systems research (for example, Prade, 1983), and the second by attributing interest groups to each rule and then using the inspection of rule usage in the review stage to provide information on the extent of interest group recognition.

There are many possibilities for improving existing features of the program; four are being incorporated in the new version. First, the output from the program can be improved. Rules, data, the activity-allocation table, the final zoning scheme, and the reasons behind the allocations will all be provided on paper output so that the planner can discuss the plan with others not present during the operation of the program. Additionally, computer graphics will be provided so that the planner can interactively inspect and adjust the activity-allocation table and map as they are formed.

Second, the syntax of the rules will be expanded to include both the allocation of predefined zonings to planning units (that is, the second model of a zoning scheme will be operationalized), and the use of more varied criteria to govern the allocation. As an example of the first of these improvements, rules such as: Assign rural 1(a) zonings to planning units where the agricultural capability rating is class 1 or 2, will be incorporated in the program.

Third, one of the central features of ADAPT is the flexible provision of information to the planner during both the construction of the activity-allocation table (stage 3) and the review of this table (stage 4). This feature will be markedly improved by allowing ADAPT to infer information from an extended knowledge base in the manner of an expert system.

Last, ADAPT will be extended to assist the planner to draw up a zoning scheme from the activity-allocation table and map (stage 5), through the same decision-making philosophy that is embedded in stage 3. Thus, rules to help the planner assess "similarity" between rows of the table, support for a dialogue between planner and computer, and a memory for reasons behind the amalgamation of rows of the table will all be incorporated in the new version of the program.

Conclusions

The ADAPT program is specifically designed to assist land-use planners in producing a zoning scheme. The program aids the planner in making decisions (rather than automating the decision-making) by recording and displaying relevant data and knowledge, by being highly interactive (both asking questions of the planner and responding to questions from the planner), by keeping a record of how and why decisions were made, and by taking simple decisions whilst leaving complex decisions to the planner. Its design rejects the notion that the typically complex decisions within the task of drawing up zoning schemes are amenable to automation.

The philosophy behind the use of the program is in accordance with that of the decision support systems now being developed for managerial purposes. Whereas DSS developers have looked to research into expert systems for techniques to represent and apply specialized knowledge, we have heuristically developed the knowledge-handling mechanisms of ADAPT in response to the characteristics of the problem domain, via the production of zoning schemes. Nevertheless many of the features of ADAPT are similar enough to those of knowledge-based programs to warrant its being classified as a knowledge-based DSS. In particular, explanations can be provided for decisions taken; the knowledge is recorded in a modular form and can be easily modified; and interactions between the system and the planner are conducted in a simple form of English.

The design of ADAPT has been user driven and not technology driven and, although it is tempting to use the techniques of knowledge-based systems to extend the capabilities of the program, these improvements will only be implemented after a number of practical applications have proven their worth. At present we would claim that the program is suited to the production of local government zoning schemes and possibly those public land zoning schemes which are complex in the sense described earlier. Experience needs to be gained with the decision-making environment of other land-use planning problems before the program can be extended to these problem areas. However, the decision-aiding approach which advocates "joint human-computer decision-making process and in which the computer supports the process by furnishing pertinent information" (Bonczek et al., 1981, page 33) is likely to remain the foundation on which such programs would be built.

ADAPT is written in FORTRAN-77 and can be run on most common 16-bit microcomputers. The compiled program requires 150K of memory; additional space is needed for mass storage of the rules, the data, and the record of the allocation. However, FORTRAN is not suited to the highly recursive nature of ADAPT, and future versions of the program will be written in a recursive symbol-manipulating language such as LISP or PROLOG. Copies of the program are available from the authors.

References

Anderson, B.F., 1981. "Multi-objective Multi-public Method of Alternatives Evaluation in Water Resources Planning", Bureau of Reclamation, Department of the Interior, Washington, D.C.

Austin, M.P., K.D. Cocks, (Eds), 1978. Land Use on the South Coast of New South Wales: A Study in Methods of Acquiring and Using Information to Analyse Regional Land Use Options, (CSIRO, Melbourne).

Barr, A., E.A. Feigenbaum, 1981. *The Handbook of Artificial Intelligence,* Vol. 1, Heuristech Press, Stanford, CA.

Barras, R., T.A. Broadbent, 1982. "Progress in Planning Volume 17," Part 2/3. *A Review of Operational Methods in Structure Planning,* Pergamon Press, Oxford.

Batey, P.W.J., M.J. Breheny, 1978. "Methods of Strategic Planning", *Town Planning Review,* 49, 502-518.

Bonczek, R.H., C.W. Holsapple, A.B. Whinston, 1981. *Foundation of Decision Support Systems,* Academic Press, New York.

Brachman, R.J., S. Amarel, C. Engelman, R.S. Engelmore, E.A. Feigenbaum, D.E. Wilkins, 1983. "What Are Expert Systems?" in *Building Expert Systems,* Eds, F. Hays-Roth, D.A. Waterman, D.B. Lenat, Addison-Wesley, Reading, MA, pp. 31-58.

Charniak, E., D. McDermott, 1985. *Introduction to Artificial Intelligence,* Addison-Wesley, Reading, MA.

Cocks, K.D., I.A. Baird, J.R. Anderson, 1983. "Application of the SIRO- PLAN Planning Method to the Cairns Section of the Great Barrier Reef Marine Park Australia," *Divisional Report 83/1,* CSIRO Division of Water and Land Resources, Canberra.

Cocks, K.D., J.R. Ive, J.R. Davis, 1986. "Developing Policy Guidelines for the Management of Public Natural Lands," *Land Use Policy* 3, 9-20.

Cohon, J.L., 1978. *Multiobjective Programming and Planning,* Academic Press, New York.

Compagnoni, P., 1984. "Environmental Planning Method," *Working Document 84/7,* CSIRO Division of Water and Land Resources, Canberra.

Compagnoni, P., 1986. "An Environmental Planning Method for Australia Jurisdictions? Comments on the SIRO-PLAN/LUPLAN Schema," *Environment and Planning B: Planning and Design* 13, 335-344.

Compagnoni, P., J. Bell, B. Lawrence, T. Soutberg, 1982. "Data Zone Mapping for Yarrowlumia Shire Land Use Plan," *Technical Memorandum 82/11*, CSIRO Division of Water and Land Resources, Canberra.

Davis, J.R., 1985a. "ADAPT -- An Aid for Devising Zoning Schemes," in *Microcomputers for Local Government Planning and Management,* Eds, P. Newton, M. Taylor, Hargreen, Melbourne, pp. 132-142.

Davis, J.R., 1985b. "Systematizing the Production of Environmental Plans: an Australian Example," *Environmental Management* 9, 443-448.

Davis, J.R., J.R. Ive, 1985. "Dungog Shire Environmental Plan: An Application of SIRO-Plan to Local Government Planning," *Divisional Report 85/2,* CSIRO Division of Water and Land Resources, Canberra.

Davis, R., J. King, 1977. "An Overview of Production Systems," *Machine Intelligence,* Volume 8, Eds, E.W. Elcock, D. Michie, (John Wiley, New York), pp. 300-332.

Faludi, A., 1985. "Flexibility in Zoning: The Case," *Australian Planner*, 23, 19-24.

Hayes-Roth, F., D.A. Waterman, D.B. Lenat, (Eds), 1983. *Building Expert Systems,* (Addison-Wesley, Reading, MA).

Hill, M.A., 1968. "A Goals Achievement Matrix for Evaluating Alternative Plans," *Journal of the American Institute of Planners,* 34, 19-29.

Ive, J. R., K.D. Cocks, 1983. "SIRO-PLAN and LUPLAN: An Australian Approach to Land-Use Planning. 2. The LUPLAN Land-Use Planning Package," *Environment and Planning B: Planning and Design,* 10 347-355.

Ive, J.R., K.D. Cocks, 1985. "Adding a Localized Adjustment Capability to the LUPLAN Land-Use Planning Package," *Environment and Planning B: Planning and Design,* 12 455-462.

Klein, H. K., R. Hirschheim, 1985. "Fundamental Issues of Decision Support Systems: A Consequentialist Perspective," *Decision Support Systems,* 1, 5-24.

Laut, P-R, 1984. "An Assessment of SIRO-PLAN for Rural Planning in New South Wales Local Government", *Diploma or Urban and Regional Planning Thesis,* University of New England, Armidale, NSW.

Lesser, V.R., L.D. Ermin, 1977. "A Retrospective View of the HEARSAY-II Architecture" in *Proceedings of the Fifth International Joint Conference on Artificial Intelligence,* (William Kaufman, Los Altos, CA), pp. 790-800.

McDonald, G.T., A.L. Brown, 1981. "A Report on the Application of a Policy Oriented Approach to Strategic Planning in Redland Shire", *Institute of Applied Social Research,* Griffith University, Nathan, Queensland.

Michie, D., (Ed.), 1982. *Introductory Readings in Expert Systems,* (Gordon and Breach, New York).

Openshaw, S., 1977. "An Evaluation and Assessment of the Use of Quantitative Techniques in Planning," *Planning Outlook,* 20-23.

Polanyi, M., 1962. *Personal Knowledge: Towards a Post-Critical Philosophy,* corrected edition (University of Chicago Press, Chicago, IL).

Post, E., 1943. "Formal Reductions of the General Combinational Problem," *American Journal of Mathematics,* 65, 197-268.

Prade, H., 1983. "A Synthetic View of Approximate Reasoning Techniques," in *Proceedings of the Eighth International Joint Conference on Artificial Intelligence,* Ed. A. Bundy (William Kaufman, Los Altos, CA), pp. 130- 136.

Simon, H.A., 1960. *The New Science of Management Decision,* (Harper and Row, New York).

Simon, H.A., 1981. *Foreward to Foundation of Decision Support Systems, R.H.* Bonczek, C.W. Holsapple, A.B. Whinston (Academic Press, New York), pp. xi-xiii.

Timmermans, H., 1984. "Decision Models for Predicting Preferences Among Multiattribute Choice Alternatives", in *Recent Developments in Spatial Data Analysis,* Eds, G. Bahrenberg, M.M. Fischer, P. Nijkamp, (Gower, Aldershot, Hants), pp. 337-354.

Weiss, S.M., C.A. Kulilowski, 1984. *A Practical Guide to Designing Expert Systems,* (Chapman and Hall, Andover, Hants).

Winston, P.H., 1984, *Artificial Intelligence,* (Addison-Wesley, Reading, MA).

Zetter, J.A., 1974, "Applications of Potential Surface Analysis to Rural Planning," *The Planner,* 60, 544-569.

CHAPTER 5

ERS: Prolog to an Expert System for Transportation Planning

Richard K. Brail

Urban and regional transportation planners have long used sophisticated computer models to assist the decision-making process. This rich tradition has included a variety of model types and computer environments. For the most part, the models were written in the computer languages of the day, chiefly FORTRAN, with some use of the newer structured languages such as Pascal and C in more recent applications. These procedural languages were used to write complex and highly sophisticated programs. In the process of developing these systems, a number of interesting algorithms, such as the shortest path through a roadway network and the calculation of roadway volumes, were designed. The "granddaddy" of the early mainframe systems was the Urban Transportation Planning System (UTPS), supported by the U.S. Department of Transportation.

The evolution of expert system approaches suggests that traditional transportation planning analysis may benefit from a new look at building applied models. This paper will explore the use of Prolog, a logic-based computer language, in developing a simple transportation model. The presentation is divided into three parts. First, we will introduce Prolog as a language. Second, we will outline the transportation problem which will be examined. Finally, we will present a simple working transportation model written in Prolog. This actual demonstration is an excellent way to see the potential, as well as the obstacles, of using such a language. Furthermore, we will be better able to assess the potential for expert system approaches to applied transportation modeling.

Prolog as a Language

The Elements of Prolog

Prolog is declarative rather than procedural language. Rather than detailing the procedures required to generate a solution, the user specifies a problem and the computer provides a set of possible solutions. One of the key concepts of

Prolog is recursion — the ability of Prolog instructions to invoke themselves in carrying out a task.

The introduction in 1986 of a number of microcomputer-based Prolog interpreters and compilers has made this language an increasingly important option to other declarative languages. The majority of Prolog compilers and interpreters follow the language conventions of Clocksin and Mellish (1987). In Prolog, a program consists of three major elements—facts, rules and questions about the facts and rules (Clocksin and Mellish,1987:1-20). Facts and rules are generally referred to as clauses. A Prolog database might contain the following facts:

```
trips_per_acre(office_land_use,500.0)
is_county(new_jersey,somerset)
```

The first fact is a clause with a predicate, "trips_per_acre," and two arguments, "office_land_use" and "500.0." This clause can be translated into the phrase "office land use generates 500 trips per acre per day." The second fact states that "Somerset is a county in New Jersey." Note that the arguments, or atoms in Prolog, are written as all lower-case letters as single words or numbers. The phrase,"office_land_use," is an atom, a lower-case non-integer constant containing no spaces. Also, "somerset," "newjersey" and "500.0" are atoms.

Rules are the driving elements of a Prolog program. Rules are the conditional "if-then" statements which lie at the base of Prolog, and of expert systems in general. In Prolog the following is a rule:

```
dense(new_brunswick)  :- many(people)
```

The ":-" symbol means if. The rule reads: "New Brunswick is dense if there are many people." Instantiation is the process in Prolog of finding an instance among the facts and rules which is supported. Prolog responds to a goal set by a user as if it were a question which has an answer, yes or no. In Prolog, a goal is an uninstantiated query. By stating a goal as a query, the Prolog interpreter or compiler is directed to instantiate the clause as a query. Prolog will answer "yes" or "no" to the query, depending on whether or not the goal was instantiated. For example, we could set up clauses containing trip generation information in terms of the number of daily trips per acre produced by different land use categories:

```
trips(office_general, 145.0).
trips(office_medical, 426.0).
trips(restaurant_quality, 200.0).
trips(restaurant_popular, 932.0).
```

We could then ask:

```
trips(office_general, X).
```

The response will be:

```
X = 145.0 .
```

Prolog will search all the clauses seeking to instantiate the land use "office _general." When instantiated, Prolog returns the associated number of trips, 145. "X" in the query is a variable to which 145 is instantiated in a successful search. Variables in Prolog are line variables in other programming languages, taking on a range of values. Variables start with capital letters, so that Population_1980 (a variable) is not equivalent to population_1980 (an atom).

The structure is a generalized term to refer to a single object which contains a set of other objects. Essentially, a structure is a compound term, or set of nested clauses, according to Clocksin and Mellish (1987:23). Facts can be nested, as is shown in this example:

```
contains(mercer_county,trenton(city, 50000))
```

This structure can be interpreted as "Trenton, in Mercer County, is a city of 50,000 inhabitants."

Using Lists

A list is an ordered sequence of indeterminate length which contains zero or more elements. Lists are central to Artificial Intelligence (AI) languages. In fact, the language LISP stands for LISt Processor. Unlike BASIC or Pascal in which an array must be declared with a fixed number of elements, a list in Prolog can be altered at any time. By convention, a list labeling system uses brackets to demarcate elements. Two examples are shown below:

```
[1,2,4,6]
[shop,school,factory]
```

Lists can be broken in two parts—the head and the tail. Prolog uses a vertical bar to separate the two pieces. Thus, in the following list:

```
[45|23,56,89]
```

the 45 is the head and the numbers 23, 56 and 89 are contained in the tail. The delineation of a list into these two parts is useful because we can then use the recursive property of Prolog to manipulate lists. We will discuss how to manipulate lists by outlining two of the common list manipulation routines often found in the literature: 1) appending a new element to a list (append), and 2) writing out a list on the screen or other output device (write_list). These routines are standard and quite useful in a variety of settings. They are not easily

understood without some understanding of Prolog programming practices, and
they demonstrate the special nature of the language.

Let us assume that the Prolog program is designed to read in a number of
land use categories and store them in a list. The program below demonstrates
the use of the append and write_list clauses.

```
================================================
% NOTE: % MEANS A COMMENT FOLLOWS
demonstrate :-

Old_array = [residential, commercial, industrial],
write('Get new land use '),
read(Landuse),nl,    %nl=new line(carriage return)
append([Landuse],Old_array,New_array),
write('The list of land uses is '), nl,
write_list(New_array).

% The following "generic" clauses are called from
% the program above.

append([],L,L).
append([X|L1],L2,[X|L3]) :- append(L1,L2,L3).

write_list([]).
write_list([X|Y]) :- write(X),nl,write_list(Y).

================================================
```

The program above is started by stating demonstrate as a goal. The
comma designates the conjunction "and." Hence, the goal demonstrate is instantiated if the Old_array list is set to contain the atoms residential, commercial and
industrial land uses and if "Get new land use" is written on the computer screen
and a variable Landuse is typed in (read) from the keyboard and if a carriage
return (nl) is sent to the screen, and so forth, down to the period after the
write_list(New_array) clause. Prolog will proceed down a set of compound
clauses as long as each succeeds. As soon as a clause does not succeed, the program will backtrack and look for another way of satisfying the goal.

In this example, the list variable, Old_array, is unified with (or assigned)
three land use categories—residential, commercial and industrial. The write
command will put the phrase "Get new land use" on the screen, and the read
command will unify the user's answer with the variable Landuse. The append
command will attach Landuse to the Old_array to produce the New_array list.
The write_list command will write the New_array list to the screen.

Generic append and write_list clauses are essential to the running of a
Prolog program containing references to these clauses. We will always need to
have the two following append clauses somewhere in the program:

```
append([],L,L)      %1st generic clause
append([X|L1],L2,[X|L3]) :- append(L1,L2,L3) %2nd generic clause
```

These generic clause sets are mandatory because they exactly define what is meant by the proposed action. The clause:

```
append([Landuse],Old_array,New_array)
```

means that the variable Landuse, entered from the keyboard, should be added to Old_array, the existing land use array, to produce New_array, the newly-formed list.

Prolog carries out the append, or joining of lists, by utilizing two common principles of central importance to the language—recursion and exit conditions. Recursion refers to a programming operation calling itself. However, there must be an exit process in a recursive situation to avoid an infinite loop (Clocksin and Mellish, 1987). The generic append clauses show both recursion and exit conditions at work. The first clause contains the exit condition—the empty list—while the second generic clause recursively calls itself to add L1 to L3.

Prolog relentlessly seeks to find a true solution to the query. Initially, Prolog will seek to instantiate the first append clause it finds. Prolog only moves on to the second clause if L1 is not empty, since the first clause will fail. An empty list is designated in the clause above with [ ]. Thus appending goes on as long as the entering list has elements. When the list, L1, is emptied, then the recursive call to the first clause will succeed and the process will stop. In the second generic append clause the head "X" of L1, the first list, becomes the head of the new list, L3, if there is an element actually in L1. This append clause will continually add successive elements from L1 to L3 until L1 is empty.

The write_list generic clauses show the recursive principle more clearly. The second generic clause:

```
write_list([X|Y]) :- write(X),nl,write_list(Y).
```

contains a call to write_list itself. The clause will write the head, X, of the list, skip to a new line, and do the write_list on the tail of the list, Y.

Developing and Using Databases

In a general sense, a Prolog program can be defined as an "intelligent" database. The program contains the logic needed to do the specified operations and the data needed for the tasks. One of the advantages of a Prolog environment for applied models is the seamless integration of data and logic. Embedded in Prolog is the potential for dynamic database manipulation in which the user can modify existing data to reflect new information and new experiences. The "self-modifying" capacity of a Prolog program, in which the program can ask the user to enter new information, has already been seen in the discussion of the append clause. A user-entered land use category into the variable Landuse

could be appended to an existing list. This binding of lists Landuse and
Old_array into New_array is temporary and evaporates as soon as the append
clauses are called again. This means that each time a new land use is entered by
the user, the earlier entry is lost both as the variable Landuse and in list
New_array.

While the use of lists in Prolog is conceptually appealing, applied model-
building usually requires the extensive use of continually manipulated and
updated data. There are two basic commands in Prolog used to enter data,
coded as clauses, into a Prolog program. These are the asserta or assertz predi-
cates. The asserta predicate will enter a clause into a database at the beginning
of the group of similar clauses, while asserts puts the clause at the end. For
example, we might want to store the average number of trips per day generated
per acre from different land use categories. The first clause below says that single family low density residential housing (single_fam_res_low) generates 9.3
trips per acre per day. The second clause contains the same information for
community-scale shopping center trips.

```
trips(single_fam_res_low,9.3).
trips(shopping_center_community,330.0).
```

A simple program to enter new trip generation information into a Prolog
program is shown below:

```
========================================

%   The goal enter asks for new land use
%     and associated trips

enter(Landuse,Number) :-
  write('Get land use  '),
  read(Landuse),nl,
  write('Number of trips'),
  read(Number),nl,
  assertz(trips(Landuse,Number)).

========================================
```

This clause will query the user for a land use category to be stored in the variable Landuse and the number of trips associated with that land use to be put into
Number. The assertz clause will enter the identified clause into a database
stored within the Prolog program as the last clause of that type. If the user
enters "office_medical" and "426.0" to the Landuse and Number queries, then
the trips clauses would read:

```
trips(single_fam_res_low,9.3).
trips(shopping_center_community,330.0).
```

```
trips(office_medical,426.0).
```

The enter goal is itself a structure with two arguments, Landuse and Number. As will be shown, these arguments are necessary in order to pass information among different structures within a program.

Beyond the seamless integration of databases with logical operations, Prolog databases contain an important advantage over the standard database structure found in other applications software. Prolog databases are made of clauses which contain both pieces of information and the relationships among the pieces. A Prolog program containing the trips clauses above would "know" that low density single family residential land use generates 9.3 trips. The relationship, trips, is stored along with the land use and the number. In a standard flat file configuration, both the land use and the trips would be stored as fields without any indication that they are connected.

The Programming Environment

There is a general distinction between interpreters and compilers of computer languages. An interpreter converts the source language into machine code one line at a time and performs the prescribed operation immediately. The BASIC language often provided with a microcomputer is an interpreter language. A compiler, on the other hand, will convert the entire program into machine instructions. Only Prolog interpreters were introduced initially. The language has a dynamic environment which permits the adding and subtracting of facts and rules at any time: this type of program alteration is best suited to an interpreter.

Prolog programs can modify themselves, contradicting the concept of a program completely compiled into machine code which cannot be changed. A number of compilers have emerged regardless (Covington and Vellino, 1986). These Prolog compilers generally will compile the original program but will run the compiled code within an interpreter so that the program can change as it is running. We will use an interpreter developed by Arity Corporation which generally follows the Prolog language conventions found in Clocksin and Mellish (1987).

An Urban Transportation Planning Example

Introduction

Transportation planning is a well-developed field with a rich set of concepts and a variety of tested methods and techniques. The field owes its current computational sophistication to the invention of the high-speed computer. Starting with the studies in the 1950's in Chicago, Detroit and Pittsburgh, experts began to model traffic flows, connecting these flows back to the land uses which generated them. The traditional transportation planning process contains four steps—trip generation, trip distribution, modal split and traffic assignment.[1] Trip generation estimates how many trips would leave and enter a defined traffic zone. Statistical connections are drawn between land use categories and trip generation rates. Trip distribution models tells us how many trips would be made between any two zones. Modal split defines the number of trips made by alternative transportation modes, such as auto or bus. Traffic assignment loads the trips onto a simulated road network. While there have been extensive efforts to improve on this basic four-step process, it is still widely used as a basis for operational models.[2]

The urban transportation planning process contains an extensive body of knowledge which is amenable to explicit rule creation and which is based on a wide variety of mathematical algorithms. Experts and expert knowledge abound, leading us to ask: what form would an expert system take for transportation planning? We will build a simple "expert response system" based on the widely used "Quick Response System."

The Quick Response System

The Quick Response System (QRS) is a simplified transportation model system developed in the 1970's. The original system was a manual method which received wide distribution. It encompassed the major components of the urban transportation planning process—generation, distribution, modal split and assignment. The Quick-Response Urban Travel Estimation Techniques and Transferrable Parameters: User's Guide manual (Sosslau et al., 1978) contained extensive sets of data, providing generalized default information for a local situations, and a variety of models.

The manual system outlined in the QRS User's Guide made great demands on the analyst, requiring serious familiarity with a hand calculator. Trip assignment was done manually, meaning that traffic volumes were calculated individually and entered into a summary table by hand. This system was designed for the smaller transportation problem. The larger ones were still assumed to be handled by transportation packages run on the mainframe computers available in the late 1970's.

Ironically, the publication of the User's Guide for QRS occurred in 1978, the same year that the Apple II microcomputer was made commercially available. The microcomputer is the perfect vehicle for QRS, because it can handle tedious hand calculations easily. Professionals at the U.S. Department of Transportation quickly recognized the potential of QRS on microcomputers and commissioned a version for the Apple II microcomputer.[3] The Apple II version was replaced by a version for IBM-compatible personal computers. These QRS packages have been written primarily in Pascal, and recent versions of QRS contain a full-blown traffic assignment model.

The Expert Response System

What distinguishes QRS from an expert system version, which we have dubbed the "Expert Response System (ERS)?" The standard QRS system contains both models and data. The models contain sets of computational algorithms which are based on a long tradition of work on how to model transportation demand. The QRS system can tell us how many trips will occur between any two traffic zones, or how many vehicles will travel on a particular highway link during a peak hour. In this sense, the models reflect the "wisdom" of experts in the field. Yet, QRS is not an expert system.

There are two significant differences between QRS and ERS. First, a well-designed Expert Response System could respond to hypothetical if-then questions about various aspects of the land use and transportation within the system. The ERS, as a Prolog program, would contain databases which store both data and the relationships among data elements. Questions can be asked of the databases in a free-form fashion. Second, the ERS would contain self-modifying databases which can be changed within the system. The ERS could learn from interactions with the user, should the user have information important to the goals established. Ideally, the ERS would be able not only to tell us the traffic volume on a roadway, but also offer reasoned judgment on the seriousness of the traffic congestion problem. The judgmental aspects of an analysis is a distinguishing characteristic of the expert system approach (Han and Kim, 1988).

In the broadest sense, expert systems codify and put within a common environment elements heretofore kept separate. Of course, it is entirely possible to do a QRS analysis which simulates an expert system approach. One could simply run QRS under a set of different assumptions, and do goal-seeking as a manual process. To answer the question, "what are the traffic impacts of a new shopping center?" the QRS system could be run under the appropriate scenarios.

We, of course, want to go further than this manual system and build an integrated expert system. A fully-fleshed out ERS would contain a broad framework for answering questions and responding to goal-seeking across a variety of users and situations. To a degree, an ERS would provide a commonality of reasoning which could be carried from one transportation problem to another. Self-modifying databases would permit the continual enriching of the Expert Response System, building on the experiences learned across applications.

The Pieces of a Simple ERS System

The simple Expert Response System proposed here is a Prolog program which would need extensive revisions to make it a useful tool in the planning field. At the simplest level, the expert system is a Prolog program which does basic traffic impact analysis. Such a model will be presented later in this chapter (Table 5.1). At the most complex level and beyond the scope of this paper, the expert system would contain an extensive urban transportation planning methodology, detailed databases, and judgmental facts based on discussions with domain experts.

The program presented here, the Basic Expert Response System (BERS), estimates the number of vehicles which will be generated by a proposed development made up of a variety of different land use categories. For example, a planned unit development might contain low and medium density residential land uses, a shopping center, and offices. BERS will estimate these impacts as they affect an adjacent roadway. Our discussion of the Prolog program is broken into four areas:

1. The "trips" database

2. The goals for the system

3. The input data section

4. The projection section

The "Trips" Database. The basic data are the trip generation rates for different land uses stored in the trips clauses at the beginning of the program. The values are daily rates per acre. Other spatial units, such as square feet of interior space, could be used depending on the data sources available. Obviously, other classes of database clauses could be stored within the program.

The Goals for the System. A Prolog program can support a variety of goals, all of which can access the same databases and logical structures. In fact, the flexibility of a Prolog environment permits "on the fly" queries. The ERS program contains two formal goals for which logical structures have been developed. The first of these is modify, which is located just below the trips clauses in the program listing. In fact, modify is listed twice, an example of the instantiation process. The modify clause will add new land uses and associated daily trips per acre to the existing trips clauses in the database.

Entering modify as a goal forces Prolog to attempt instantiation. ERS will look at the first modify clause, which calls the stopif(Done) clause. This clause will succeed if the user types "yes" to the query "If finished entering new land uses, type 'yes.'" If the user does not type "yes," Prolog will go on the second modify clause, seeking instantiation there. This second clause asks the user for the land use and the number of trips. At the end of the second modify clause is the call to itself, modify. This is a clear example of recursion. The exit condition from this potentially infinite recursion process is the response "yes" to the question asking if we are finished entering data.

The second formal goal, run, does the actual calculations of traffic impacts. The run clause calls two other clauses:

```
run :-
  get_inputs(Volume,Capacity,Cur_split,Pop_pct_change),
  do_projection(Volume,Capacity,Cur_split,Pop_pct_change).
```

The get_inputs and do_projections clauses have sets of arguments (Volume,

Capacity, etc.) which permit the passing of information among the clauses. The get_info clause will query the user about relevant information, while do_projection does the calculations and writes the output to the computer screen.

The Input Data Section. The program will query the user about three kinds of data—data on the proposed development (get_development_info), information on current traffic conditions at the site (get_traffic_info), and background data on population growth (get_population_info).

```
get_inputs(Volume,Capacity,Cur_split,Pop_pct_change) :-
   get_development_info,
   get_traffic_info(Volume,Capacity,Cur_split),
   get_population_info(Pop_pct_change).
```

The get_development_info clause queries the user for the type and magnitude of the land uses planned in the development. There are two elements in this clause which are of particular interest. First, basic Prolog handles loops awkwardly. The Arity Corporation version has added ctr_set and ctr_inc, two clauses which act as a looping counter. The following clauses will establish a counter "0" and increment from 1 to Y:

```
ctr_set(0,1), repeat,
ctr_inc(0,Y),
```

The repeat clause is always true, and will keep ctr_inc incrementing until Y is reached (Clocksin and Mellish, 1987:121-123). The variable Y is the number of land use types which are to be included in the development.

Second, the getlu and putlu clauses (see Table 5.1) query for a land use and then store the answer in a daily clause using appendz. Of particular note in putlu is the relationship among the compound clauses:

```
putlu(Landuse,Acres) :-
    trips(Landuse,Trips),
    Daily_trips is (Acres * Trips),
    assertz(daily(Daily_trips)).
```

The trips (Landuse,Trip) clause will attempt to instantiate the land use entered in the getlu clause to the trips database. If the land use in the development matches a land use in the database, then trips(Landuse,Trips) is instantiated and Prolog moves on to the next clause. The "is" in Daily_trips is (Acres * Trips) indicates an equation: Daily_trips is instantiated to the result of multiplying Acres and Trips.

There are extensions to our simple program at which Prolog would be adept. For example, by specifying a set of rules on searches, the program could offer a set of known land use categories as options for one which has an unknown trips per acre parameter. For example, the user might want to find the trips per acre for a commercial enterprise, an auto supply store. By specifying the type of commercial activity, perhaps stating that the land use is similar to, say, a plumbing supply house, the program could "guess" at the trip generation rate.

The get_traffic_info clause inquires about traffic information. The program will also ask the user about background population growth in the get_population_info clause. This data creates a basis for increasing background traffic levels under the assumption that regional population growth will mean additional traffic volume. The get_population_info clause calls check_pop_data, which here only calculates a simple population percent change. It would be possible to extend check_pop_data to query a range of demographic data as a basis for a more comprehensive projection of background traffic growth.

The Projection Section. The standard transportation planning model systems, such as QRS, are well developed and do a reasonable job of calculating traffic impacts across a wide variety of situations. In this example, the traffic model is simplistic, but contains a range of Prolog-based elements. The do_projection clause uses both lists and clauses. The findall clause is standard Clocksin and Mellish (1987:156-158):

```
findall(Trips,daily(Trips),L),
```

The clause will create a list L of all trips generated by the development. This list is drawn from the daily clause database as created in putlu (Table 5.1). The list of all trips is summed using a call to the sumlist clause which adds the elements of list L and stores the summation in Sum_trips. Also, the number of elements in the list is calculated and stored in N:

```
sumlist(L,Sum_trips,N),
```

As should be obvious from the general clauses which make up a sumlist at the end of the program, adding a set of numbers is not as simple as in a procedural language, such as BASIC or Pascal. In fact, arithmetic operations in Prolog are weak. Some Prolog implementations offer the possibility of using programs written in another language within the general environment. For example, a mathematical subroutine in FORTRAN, Pascal, or C could be called by Prolog in some vendor implementations.

The remaining part of the do_projection clause does a set of calculations and prints out the results. The trip generation calculations are done on a daily basis and converted to peak hour trips (Peak_hr_trips). The existing traffic on the road is grown by the expected population percent increase in the study area between the current and projection years. The variable Proj_volume is instantiated as this traffic level. The traffic from the development is assumed to exit on one roadway, and is split between left and right turns for both study area traffic and for the proposed development. The final output (Tot_right_trips and Tot_left_trips) sums the study area and development traffic.

Table 5.1. Prolog source program listing for the Expert Response System.

```
%=======================================
% THE TRIPS PER DAY PER ACRE FILE
%=======================================
trips(single_fam_res_low,9.3).
trips(single_family_res_medium,40.8).
trips(single_family_res_high,70.0).
trips(multiple_family_res_low,150.0).
trips(multiple_family_res_medium,210.0).
trips(shopping_center_regional_large,580.0).
trips(shopping_center_regional_medium,370.0).
trips(shopping_center_community,330.0).
trips(shopping_center_neighborhood,660.0).
trips(office_general,145.0).
trips(office_medical,426.0).
trips(restaurant_quality,200.0).
trips(restaurant_popular,932.0).
trips(restaurant_fast_food,1825.0).
trips(recreation_commercial,296.3).
trips(park,60.0).
trips(hospital,40.0).
trips(industry,59.9).

%=======================================
% THE MODIFY RULE
%=======================================

modify :-
   stopif(Done).

modify :-
   write('Name of land use to be added '),nl,
   read(Landuse),nl,
   write('Number of daily trips per acre for '),
   write(Landuse),nl,
   read(Number),
   assertz(trips(Landuse,Number)),
   modify.

%=======================================
% THE RUN RULE
%=======================================

%== The "run" rule below contains the model.==

run :-
   get_inputs(Volume,Capacity,Cur_split,Pop_pct_change),
   do_projection(Volume,Capacity,Cur_split,Pop_pct_change).
```

Table 5.1. (continued).

```
%==========================================
% GET INPUTS
%==========================================

get_inputs(Volume,Capacity,Cur_split,Pop_pct_change) :-
  get_development_info,
  get_traffic_info(Volume,Capacity,Cur_split),
  get_population_info(Pop_pct_change).

get_development_info :-
  write('DEVELOPMENT INFORMATION;'),nl,
  write('Number of different land use categories: '),nl,
  read(How_many),nl,nl,
  ctr_set(0,1), repeat,
  ctr_inc(0,Y),
  getlu(Landuse,Acres),
  putlu(Landuse,Acres),
  Y==How_many.

  getlu(Landuse,Acres) :-
   write('Landuse to be developed:  '),
   read(Landuse),nl,
   write('Number of acres: '),
   write(Landuse),nl,
   write(' to be developed  '),
   read(Acres),nl,nl.

  putlu(Landuse,Acres) :-
   trips(Landuse,Trips),
   Daily_trips is (Acres * Trips),
   assertz(daily(Daily_trips)).

get_traffic_info(Volume,Capacity,Cur_split) :-
   write('INFORMATION ON ROADWAY:'),nl,
   write('Hourly design capacity of roadway at the site: '),
   read(Capacity),nl,
   write('Most recent highest hourly volume: '),
   read(Volume),nl,
   write('Current one-way percent split of total'),nl,
   write('  volume traveling to the right in front'),nl,
   write('  of the development: '),
   read(Cur_split),nl.

get_population_info(Pop_pct_change) :-
   write('INFORMATION ON STUDY AREA:'),nl,
   write('Name of the study area: '),
   read(Area_name),nl,
   write('The current population of '),nl,
   write(Area_name),write(': '),
   read(Cur_pop),nl,
   write('The population projection in target year for'),nl,
   write(Area_name),write(': '),
   read(Fut_pop),nl,
   check_pop_data(Cur_pop,Fut_pop,Pop_pct_change).
```

Table 5.1. (continued).

```
check_pop_data(Cur_pop,Fut_pop,Pop_pct_change) :-
  Pop_pct_change is ((Fut_pop/Cur_pop)-1.0),
  write('The projected growth in population is'),
  write(' expected to be '),
  write(Pop_pct_change),nl,nl.

%========================================
% DO PROJECTION
%========================================

do_projection(Volume,Capacity,Cur_split,Pop_pct_change) :-
  write('THE PROJECTION:'),nl,
  write('Peak hour % of total daily trips from project: '),

  read(Peak_pct),nl,
  findall(Trips,daily(Trips),L),
  sumlist(L,Sum_trips,N),

  write('Total daily trips from the development: '),
  write(Sum_trips),nl,

  Peak_hr_trips is (Peak_pct/100.0) * Sum_trips,

  write('Peak hour trips: '),
  write(Peak_hr_trips),nl,

  Proj_volume is Volume * (1.0 + Pop_pct_change),

  write('Projected total roadway volume: '),
  write(Proj_volume),nl,
  write('The projected one-way percent split out'),nl,
  write('  of the development turning right: '),

  read(Dev_split),nl,nl,

  Proj_right_trips is Proj_volume * (Cur_split/100.0),
  Proj_left_trips is  Proj_volume - Proj_right_trips,

  Dev_right_trips is Peak_hr_trips * (Dev_split/100.0),
  Dev_left_trips is  Peak_hr_trips - Dev_right_trips,

  Tot_right_trips is Proj_right_trips + Dev_right_trips,
  Tot_left_trips is Proj_left_trips + Dev_left_trips,

  write('The total No. of trips traveling to the right: '),

  write(Tot_right_trips),nl,
  write('The total No. of trips traveling to the left: '),
  write(Tot_left_trips),nl.

  do_projection(Volume,Capacity,Cur_split,Pop_pct_change).
```

Table 5.1. (continued).

```
%==========================================
%  GENERAL CLAUSES
%==========================================
stopif(Done)  :-
write('If finished entering new land uses, type "yes."'),
    read(Done),
    Done == 'yes'.

sumlist([],0,0).
sumlist([H|T],Sum,N)  :- sumlist(T,S1,N1),
Sum is H+S1,
N is 1+N1.

write_lis([]).
write_list([X|Y])  :- write(X),nl,
write_list(Y).
```

Conclusions

What have we learned from this excursion into Prolog for transportation planning? First, Prolog is strong on integrating databases and the logic needed to operate on these databases. Databases do not need to be static entities derived from earlier studies, but rather can change "on the fly" depending on the situation. Expert systems in transportation planning will be heavily data dependent. To the degree an expert system can integrate both more recent data and judgement calls with well-developed models, an ERS could be a powerful analytic tool.

Second, Prolog is weak computationally. The mathematical sophistication often needed for transportation models is not available in Prolog. The best example is the sumlist example in which a set of clauses is needed just to add up trip generation numbers. The need to access models written in another language is central to the use of Prolog in transportation analysis. Of course, this need to access models written in other languages holds for any expert system shell or language designed for applied modeling.

Third, Prolog, as currently constituted, does not deal with uncertainty and confidence intervals. This may be a fatal flaw affecting Prolog's consideration as a serious vehicle for expert system development. Prolog uses a two-value logic, "succeed" and "fail," to solve for a goal. It is unclear how Prolog could be altered from saying "yes" to saying "highly likely" or "unlikely." Also, Prolog cannot handle confidence intervals, in which a range is placed about a point estimate (e.g., 1,000 trips plus or minus 250).

Finally, while Prolog is not a complete development environment for expert system creation in transportation planning, the language is a powerful addition to our understanding of goal-directed modeling. Any expert system development tool for transportation planning, whether it be a language like Prolog or a software shell, will have to recognize the need to do extensive computations and to maintain and access databases.

Notes

1. There are a number of texts which outline the urban transportation planning process (Black, 1981; Stopher and Meyburg, 1975). A fairly recent text by Meyer and Miller (1984), while less mathematical, contains broad discussion of alternative model structures.

2. There have been a broad number of criticisms of the classic four-step transportation planning process (Meyer and Miller, 1984). Choice theory offers an option to the traditional demand analysis approach. While beyond the scope of the investigations here, it would be exciting to introduce choice theory concepts into an expert system design.

3. The University of California, San Diego (UCSD) p-System was developed at the University for the Apple II as a powerful alternative to the primitive operating system distributed by the Apple Computer Company. The p-System stood for pseudo-system, a reference to the creation of an intermediate language between an English-like programming language and the machine language which is understood by the microprocessor. The concept has its roots in the work of Nicholas Wirth in Switzerland who invented a highly structured programming language, Pascal. Wirth thought the pseudo-system would solve the problem of running the same programming language on different microprocessors with different machine language codes. Hence, the Pascal compiler of Wirth created "pseudo-code," a generalized code close to machine language. This pseudo-code would be converted to a particular microprocessor's machine code by a "run-time interpreter." Jensen of the University of California, San Diego popularized the p-System and helped establish Pascal as a microcomputer language. The original version of the QRS system was written in Pascal for the Apple II. When the IBM personal computers became popular, QRS was ported over to run under the p-System. However, the cost and slowness of the p-System were negative factors, and the QRS system was converted to run under the standard IBM operating system. However, QRS is still written mostly in Pascal.

References

Black, John, 1981. *Urban Transport Planning,* Baltimore: Johns Hopkins University Press.

Clocksin, W.F., and C. S. Mellish, 1987. *Programming in Prolog,* Third, Revised and Extended Edition. New York: Springer-Verlag.

Covington, Michael, and Andre Vellino, 1986. "Prolog Arrives," *PC Tech Journal,* 4:11 (November), 52-67.

Han, Sang-Yun, and T. John Kim, 1988. "Intelligent Urban Information Systems: Review and Prospects," *Planning Papers,* Number 88-04. Urbana, Illinois: Department of Urban and Regional Planning, University of Illinois at Urbana-Champaign.

Meyer, Michael D., and Eric J. Miller, 1984. *Urban Transportation Planning: A Decision-Making Approach,* New York: McGraw-Hill.

Sosslau, Arthur, Amin B. Hassam, Maurice M. Carter, and George V. Wickstrom, 1978. Quick-Response Urban Travel Estimation Techniques and Transferrable Parameters: User's Guide. *National Cooperative Highway Research Program Report 187,* Washington: Transportation Research Board.

Stopher, Peter and Arnim H. Meyburg, 1975. *Urban Transportation Modeling and Planning,* Lexington, Massachusetts: Lexington Books.

CHAPTER 6

RTMAS: An Expert System for Real Time Monitoring and Analysis of Traffic During Evacuations

Frank Southworth, Shih-Miao Chin
and Paul Der-Ming Cheng

An imminent hurricane, a problem with a nuclear reactor, the escape of a dangerous gas, a large chemical spill, or even a serious threat from a foreign country are all examples of some very recent reasons for large urban populations to leave their homes until the threat to their lives and property has passed. Whether largely spontaneous, or carefully orchestrated by local authorities, such evacuations may take place over many hours or even days. To date, there has been no significant monitoring of large scale evacuations, since they occur as either rapidly developing localized threats, such as hurricanes, or for more protracted threat build-ups.

The RTMAS system is being developed to continuously monitor and cumulate the net outflows from evacuations, whatever the rate of regional exodus. In the case of the more protracted forms of evacuation, such as a nuclear threat from a foreign power, seemingly "invisible" evacuations could create otherwise unnoticeable changes in the daily traffic volume profiles of major highways. This sort of behavior was evident in the gradual, and at the time unknown, exodus of one quarter of London's population during the build up to World War II. RTMAS will allow regional emergency management planners to identify such population movements as early as possible, and provide real time monitoring and control of such evacuations.

The decision to adopt an expert system approach results from a need to deal with a potentially large number of unusual hourly or daily traffic counts, or data "outliers", due to a number of factors or combination of factors. Examples include holidays, major accidents, large scale recreational events, and even traffic counter malfunctions. While recent improvements in time series methods go a long way in identifying and accommodating certain types of data outliers within forecasted trends, they cannot interpret the causes behind such outliers. Nor can the forecasting model alone make an automatic selection of the most appropriate series of past traffic counts upon which to base the forecast series. The larger shocks introduced into the analysis by unusual traffic days or multi-hour periods associated with major regional events suggest a rule based approach to data selection and interpretation.

The rest of the paper proceeds as follows. Sections 2 and 3 provide a brief description of the current and projected RTMAS hardware and software. Section 4 provides a brief qualitative description of the nature and capabilities of current state-of-the-art time series forecasting in the presence of anomalous data outliers (for our purposes, anomalous traffic counts). This paper will not, how-ever, describe such time series procedures in detail. Section 5 describes the use of such modeling within a rule based expert system geared toward identifying and interpreting unusual traffic events not amenable in statistical analysis. Sec-tion 6 provides an example of the sort of unusual event combinations expected when analyzing traffic profiles and demonstrates how to model it. It also shows what appears to be the first telemetric recording of the traffic build-up during a hurricane induced evacuation.

In Section 7 we speculate upon further uses of a rule based expert system approach. Linking RTMAS to a network evacuation simulation model, for instance, will provide the most current information for predicting the location and timing of traffic congestion hours ahead of its occurrence. While the reader's attention is focused here upon the system's application to large scale population evacuations, it will become clear that RTMAS possesses the poten-tial to be applied to a wide variety of traffic monitoring and control situations. This potential to become a more interpretive tool is greatly enhanced by its development through expert system software.

System Hardware

Figure 6.1 shows RTMAS hardware and communications. Actual traffic counts are presently brought via modem and conventional telephone lines into either an IBM/PC-AT or a "386" architecture IBM PS2/Model 80 microcomputer using direct dial up to each traffic counter in turn.

While some telephone lines may go down during the high winds gen-erated by a hurricane, this will typically only be the case well into an evacua-tion, and only for a limited distance inland. Insulation of counter-to-communication line connections can be used to keep out moisture in non-inundated areas. Counters are reportedly otherwise very reliable machines. Accessing the telemetered counters operated by some state transportation or highways departments in the future may require work to pull such data from a regional computer which itself collects this data (as described in the feasibility

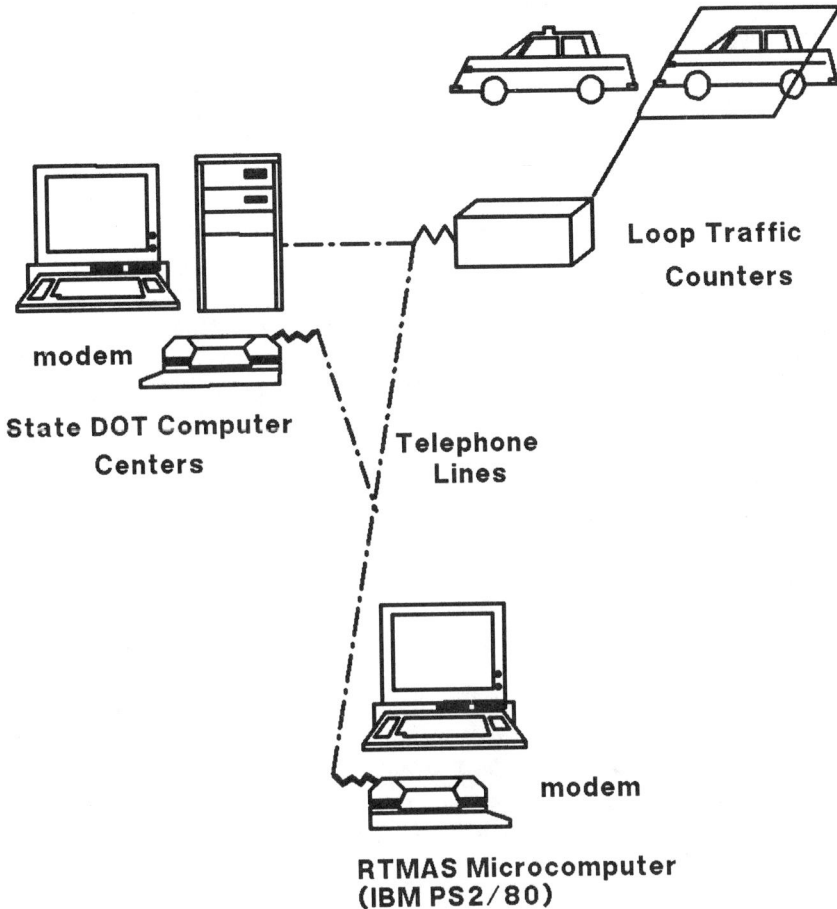

Figure 6.1. RTMAS hardware and communications configuration.

study by Southworth and Chin, 1986).

System Software

The current version of RTMA (shown schematically as Figure 6.2) integrates four major software components: GURU, AUTOBOX, CROSSTALK and a TRAFFIC PATTERN RECOGNITION PROGRAM. A fifth component, a network traffic equilibrium model that simulates the selection of evacuation routes, is currently under development as is a mouse driven network mapping and city/counter selection interface. Micro Data Base Systems Inc.'s (1987) GURU acts as the integrating software. GURU is an expert system development environment that supports a broad range of integrated knowledge processing capabilities, including rule based expert system creation and consultation, relational data base management, spreadsheet analysis, color graphics, and natural language capabilities. In addition GURU grants the ability to provide remote communication. Integration of these functions within a single software environment permits the construction of a rule based expert system with ready access to multiple sources of real time traffic data, using a variety of ways to report and, most importantly, "learn from" this new information.

Through menu screens developed in GURU, the user accesses the traffic counter communications and data retrieval programs using Microstuf's CROSSTALK. This count data, plus graphics-enhanced data series selection capabilities, are entered via easy to interpret selection menus into ASF's AUTO-BOX (ASF, 1984) time-series software, and into a TRAFFIC PATTERN RECOGNITION PROGRAM based upon principal components analysis. (Neither this principal components analysis feature nor the details of the telecommunications software are discussed in the present paper: see Southworth, Chin, and Cheng, 1987 for details).

Operationally, the GURU expert system environment, or shell, provides the user with a blank sheet on which to input his or her rules for system operation (the way data is input into one of the popular spreadsheet programs which then takes care of data management). Rules within GURU can be built using certainty factor algebra in conjunction with very flexible forward, backward, and "mixed" forward and backward chaining of rules: This allows the user to specify and combine uncertainties in a variety of ways.

Using this structure, GURU's inferencing capabilities can then be employed to interpret a traffic event with an accompanying level of certainty. For example, a rule may be designed to take the minimum of two single event certainties when calculating the certainty of the combined event occurring; or it may take their product, average, or other weighting of the two individual certainty factors. Given a suitable structuring of such rules, by which the programmer fixes the order(s) in which some rules are applied before others, highway and region-specific traffic events can be retrieved from the system's relational database and subjected to rule by rule analysis.

Both the multi-color graphics and ease with which many other software packages can be brought in under GURU's programming control were major factors in its selection for this work. Color proves particularly useful when superimposing many different days' traffic profiles on top of one another. For example, RTMAS can be asked to plot the hourly counts for every Wednesday in the year: a procedure that determines if a region's traffic counters have been properly programmed to recognize a switch to or from Daylight Saving's Time during the Spring and Fall each year.

The major drawback to current system performance is speed of response when accessing and using large amounts of historical time series data. This is most notable when cumulating complete daily traffic count totals over a full year (e.g. every Wednesday in 1987). This drawback led to the upgrade to an IBM PS2 for as the system's microcomputer. This apart, we have found GURU to offer some very powerful features well suited to a more in-depth analysis of

GURU:
EXPERT SYSTEM RULES,
RELATIONAL DATA BASE,
HISTORICAL DATA STORAGE,
NET EVACUATION TOTALS,
COLOR GRAPHICS DISPLAYS

CROSSTALK:
TELEPHONE
COMMUNICATIONS
FOR DATA TRANSFER

COUNTER
LOCATION
MAPPING
& SELECTION
SOFTWARE

AUTOBOX:
BOX-JENKINS TIME
SERIES MODELING

PATTERN
RECOGNITION
SOFTWARE:
PRINCIPAL COMPONENTS
ANALYSIS OF TEMPORAL
TRAFFIC PEAKING

EVACUATION
SIMULATION
MODEL:
DYNAMIC TRAFFIC
ROUTE ASSIGNMENT
AND SHELTER
SELECTION MODEL

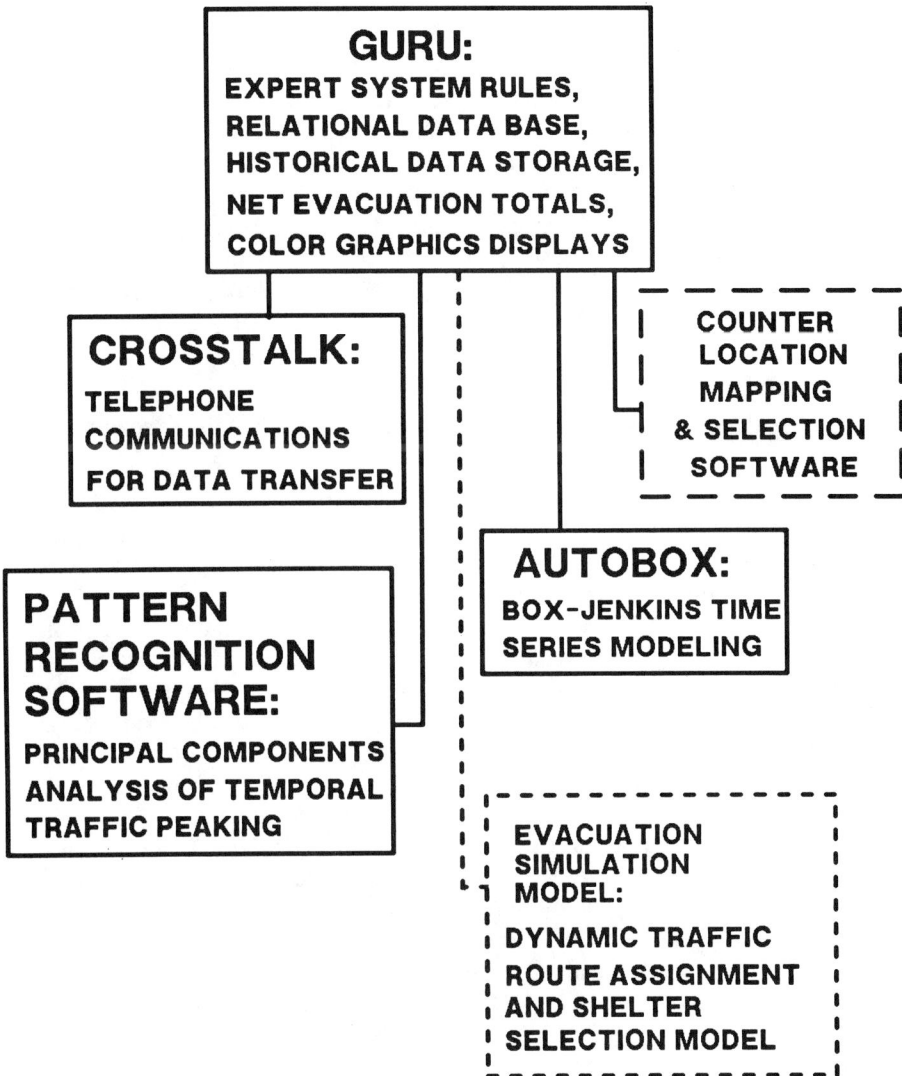

Figure 6.2. Schematic Representation of RTMAS.

traffic profiles and unusual event causation. To achieve good results, however, the programmer must invest time to learn GURU's programming language. Such an investment is similar to that required to learn any typical high level programming language and therefore, depends somewhat upon programming experience. Equal effort again is required, however, to become proficient in the use of GURU's expert system logic.

In practice, RTMAS intended for use in two types of settings: (i) the planner/engineer wishing to check current system behavior, and possibly act upon it, and (ii) the planner/engineer wishing to add new information to the system, such as adding additional traffic analysis and response rules. This second type of user has come into existence directly as a result of expert system technology. They are the experts needed to build RTMAS. As with type (i) users, type (ii) users will benefit from easy-to-understand directions for system use. The functions of RTMAS should be "transparent" to both these types of users. The system's response to the users' input should be immediately clear to the user, without the user needing to know calculation details required to produce this response. GURU was selected as the RTMAS development environment specifically with the type (ii) user in mind.

Box-Jenkins Time Series Analysis of Traffic Counts

With the introduction of telemetrically monitored traffic counters, providing a stream of real time or near real time traffic count data to the user, time series modeling has seen numerous applications in traffic forecasting. Applications range from the minute by minute forecasting of traffic adjustments to localized freeway traffic incidents (Ahmed and Cook, 1979, 1982; Davis and Nihan, 1980), to the forecasting of monthly (Nihan and Holmesland, 1980) and yearly (Benjamin, 1986) trends in traffic growth. While the RTMAS system can be adapted to receive data at whatever time interval is transmitted, the focus of this current effort is on analysis of hourly and daily traffic count data.

There are two basic approaches currently available for dealing with unusual counts, or as they are generally termed, "interventions" within time series data. The first is an extension of the Box Jenkins Autoregressive Integrated Moving Average (ARIMA) methodology, developed by, among others, Box and Tiao (1975), and by Bell (1983), and Chang and Tiao (1983), Tiao (1986), and Tsay (1986). The second approach (see Davis and Nihan, 1984) uses a Kalman filter (Kalman, 1960) to estimate linear regression models with auto regressive moving average (ARMA) errors. RTMAS uses a version of the first approach, by incorporating the AUTOBOX ARIMA software (see Reilly, 1984).

AUTOBOX provides RTMAS with the ability to model data outliers through what is commonly termed "intervention modeling", or "impact assessment modeling" in the social sciences. In non-mathematical terms, AUTOBOX reproduces the observed time series of traffic counts by modeling the series as not one but as a number of separate time series. This means fitting an ARIMA model to get an underlying time series model which picks up the major repetitive trends in the observed time series. "Residuals" are then calculated as the difference between observed and model estimated individual traffic counts. A series of multiple regressions are then carried out, using these residual counts as the dependent variable. Each regression identifies one or more unusual counts or count series within the observed series as introducing an "intervention" into the original observed series of counts. Results output from AUTOBOX include the model's maximum likelihood parameters, goodness of fit statistics, and user selected confidence intervals around each model generated traffic count.

Upon selection of either historical or real time (current day) traffic data, RTMAS allows the user to fit a particular time-series model using the selected previous and current days' hourly count data. This selection is made by counter station number, direction, and date.

For the Tampa Bay, Florida and Atlanta, Georgia systems currently operational, RTMAS has the ability to update traffic counts and model them every hour. For freeway surveillance based telemetric counting systems, such as those used on the access and exit ramps of cities such as Seattle, Chicago, and Los Angeles, this could mean updating information every few minutes. ARIMA model selection is carried out automatically within AUTOBOX for the user, who simply selects the historical data upon which to fit the model, and sends it via GURU to AUTOBOX. The user does not, therefore, need to understand the complexities of the time series modeling process. The user does, however, need to know how to interpret the AUTOBOX output. To facilitate and enhance this understanding, GURU color graphics are used.

An Expert System to Deal with Unusual Traffic Events

While the above approach can handle a wide variety of unusual count series within a clearly recognizable underlying trend, such as a typical daily traffic profile, it cannot be used to handle significant disruptions in the data series in any generically efficient manner. Nor can it be used to infer causality since, as Davis and Nihan (1984, page 432) note:

> "Even if a statistically significant effect is observed coincident with an intervention, it remains possible that some external factor or combination of factors has actually caused this effect."

The mixture of both unusually high or low counts and missing data within a time series can produce results that may be particularly difficult to interpret. Loss of a full day's data due to counter failure calls for its exclusion from the data series. Similarly, a major multi-hour delay due to roadwork or an accident is best handled by removal of this information for forecasting purposes. An effective interpretive system will be able to distinguish between those events to be excluded and those to accept and model as part of the actual variability within the data. It will take full advantage of the very effective statistical modeling provided by AUTOBOX, but place its application within a sensible context. This suggests a rule based expert system with access to (i) a memory of past, similar events, (ii) a user developed knowledge of how to select and apply time series data, and with (iii) an ability to infer causality by applying its rules to this past evidence.

For example, if the traffic count on a particular highway in a specific direction has been outside the 95% confidence interval for the last n hours (based upon Box-Jenkins time-series forecasting), then it should be brought to the system operator's attention. In practice, before any conclusion is reached, a similar premise needs to be applied over either the full set or a large sub-set of the region's counters. This is precisely the sort of rule now being constructed within RTMAS. In order to handle the many different types of events that can affect a highways' hourly or daily traffic flow profiles, a large number of such rules, and an order in which to apply them, are needed.

At least one traffic flow expert is required in the early stages of system development. The system will also need to be told the cause of each unusual traffic event by the system programmer/user. This cause can then be stored and called up for comparison purposes when a similar count profile develops, or when the same traffic date recurs. In this way the system can eventually, with appropriately defined rules, "learn" to analyze an increasingly large and varied set of traffic events without the need for user interpretation. It may also be programmed to initiate action on the basis of its findings, using a basic IF...THEN logic.

The GURU software provides the capabilities to transform the currently operational telecommunications and time series modeling-based RTMAS system into the one described here. Given that RTMAS can accurately model the hour by hour traffic count profile for most days in the year by using data from the same day in the past five weeks (i.e. to model the coming Thursday we use data from the past five Thursdays) a reasonably well defined limit on the amount of data typically needed to predict the current day's traffic profile exists. But what if one of those five days was unusual in some sense because of a holiday, major accident, road maintenance, or the counter was broken? The RTMAS system can currently recall up to the previous twelve days of traffic count data (assuming one count per hour, this provides 12x24 = 488 prior unidirectional traffic counts per station). This allows us to avoid a number of aberrant days' data. One purpose behind building within an expert system environment is to allow the system itself to recognize such aberrant days and to omit them from the subsequent Box-Jenkins time series modeling procedure.

Both historical and real time (meaning here current day) data is called by RTMAS into its data base management system. Assuming that we only "switch on" the system under some suggestion of a developing threat (RTMAS can be made to operate continuously, if desired), the first signs of unusual systemwide traffic movement will be sent to a stored history of special and seasonal traffic events. Failure to recognize the typical characteristics of such events in the current pattern leads the program to inquire about previous data on evacuation build-ups, again consulting any accessible historical evacuation data. In other words, the system is given the premise that something unusual is happening and concludes that it should examine this past evacuation data. If sufficient similarity exists between this and past spontaneous evacuation build-ups, an appropriate warning can be sent to the user. If RTMAS cannot conclude the nature of an event with a reasonably high certainty, the user should be informed and asked for further expert advice. Once the system has made a decision with a high degree of certainty as to an event's nature, it will so inform the user, providing conclusion and level of confidence in its finding.

During such system development we can expect numerous erroneous conclusions in the early stages. In terms of correctly identifying a particular event, such as a developing evacuation, there are two types of errors the system could make. A "Type I" error misinterprets the data and fails to identify the event until well into the process. A "type II" error warns the operator that an event is developing, when in reality no such event is taking place. For example, RTMAS may suggest that a particular peaking of traffic on Interstate 4 reflects a regularly scheduled sporting event, which the operator knows to be untrue on the given day. It is then the job of the operator to inform the system of this "Type II" error, if possible find out the cause of the anomalous data, and have RTMAS add this lesson to its current historical data bank for that particular highway, date and day of the week. This places a burden on the system's developers and subsequent operators to fully research and inform a system such as RTMAS of event causality.

Since no two urban systems, indeed no two highways, will have identical traffic event histories, two levels of rule development are envisioned for a fully developed RTMAS. Level One may be referred to as the generic traffic rule set, in which RTMAS is provided the fundamental knowledge with which to recognize anomalous events from traffic count data. Level Two then works within Level One rules to add highway and urban system specific information of local event dates and associated traffic patterns. This Level Two effort is best carried out by resident traffic engineers and planners, placing a burden on system developers to build rules that cover a wide range of possibilities (or, better yet, to provide easy to follow instructions for adding to or adapting current rule sets to better reflect local need). Some use of the natural language features within GURU may help the user here, another benefit of using GURU-like "integrated software".

Eventually, a well designed, locally tailored system will acquire enough highway and region specific knowledge (i.e. data and rules for data use) to pay back this effort many times. Currently, the development of Level One rules within RTMAS is being researched.

An Example of an Unusual Combination of Traffic Events

Figure 6.3 shows the location of the fourteen traffic counters in and around the hurricane prone Tampa Bay area. These counters are part of the Florida Department of Transportation's telemetric traffic monitoring system. The counters were not positioned with evacuation studies in mind, and a number of additional counting stations would be beneficial, and indeed necessary to the system's performance. Over a year, each highway in the system records a significant number of annually recurring events in the form of clearly anomalous traffic count profiles (based upon analyses of 1985, 1986, and 1987 count data). For example, Figure 6.4 shows a day by day plot of the traffic counts by counter number 123, on Interstate 275 north-bound, for each Saturday (top left diagram), and each Sunday, Monday, and Wednesday respectively in 1985. Significant annual holidays such as Labor Day and Memorial Day (both falling on a Monday) are evident from this view of the data, as depicted in the left bottom diagram. Automatically (i.e. without the trouble involved in user intervention) removing such days' count data when fitting our time-series forecasting models would prove extremely useful.

In addition to these typical holidays, each highway and cumulative urban area is likely to have its own unique traffic count profile as a result of specific regional activities. Tampa Bay traffic counts, for example, reflect the different tourist seasons associated with that part of the Gulf Coast, such as Busch Gardens and Orlando's hugely popular Disney World complex, some three hours drive from Tampa. Given the operation of counters over a number of years, RTMAS will eventually also be able to model these unique dates' traffic profiles. For example, the system will be able to estimate the likely traffic profiles associated with each July 4th, taking into account annual traffic growth on that particular highway through the use of automatically identified and calibrated multiplicative Box-Jenkins models. Such a capability will prove especially useful when, as can be expected on occasion, unusual events such as a hurricane take place during one or more of these atypical days.

Figure 6.5 shows just such an occurrence. The plots represent the eastbound traffic counts, for Station 106 on Interstate 4 (recall Figure 6.3) on the last two Fridays in August 1985 (8/23/85 and 8/30/85), along with a profile of the net difference between the two counts. Clearly evident is the unusually high net gain in outgoing traffic late on August 30. This was the approximate time hurricane Elena began to offer a threat and also (we are told) around the time that many die-hard Tampa Bay Buccaneer fans were returning home after losing to the Washington Redskins football team. In addition, at least some of the area's residents had already adjusted their usual Friday travel habits to accommodate the beginning of a Labor Day weekend (recall the Monday plots in Figure 6.4).

The system supports this plot with the data shown in Figure 6.6 (a GURU screen display). The top row of the display shows the hour in the day (t01, t02, etc.), rows two and three show the traffic counts for, respectively, August 30 and August 23, and row four the difference in the row two minus row three figures.

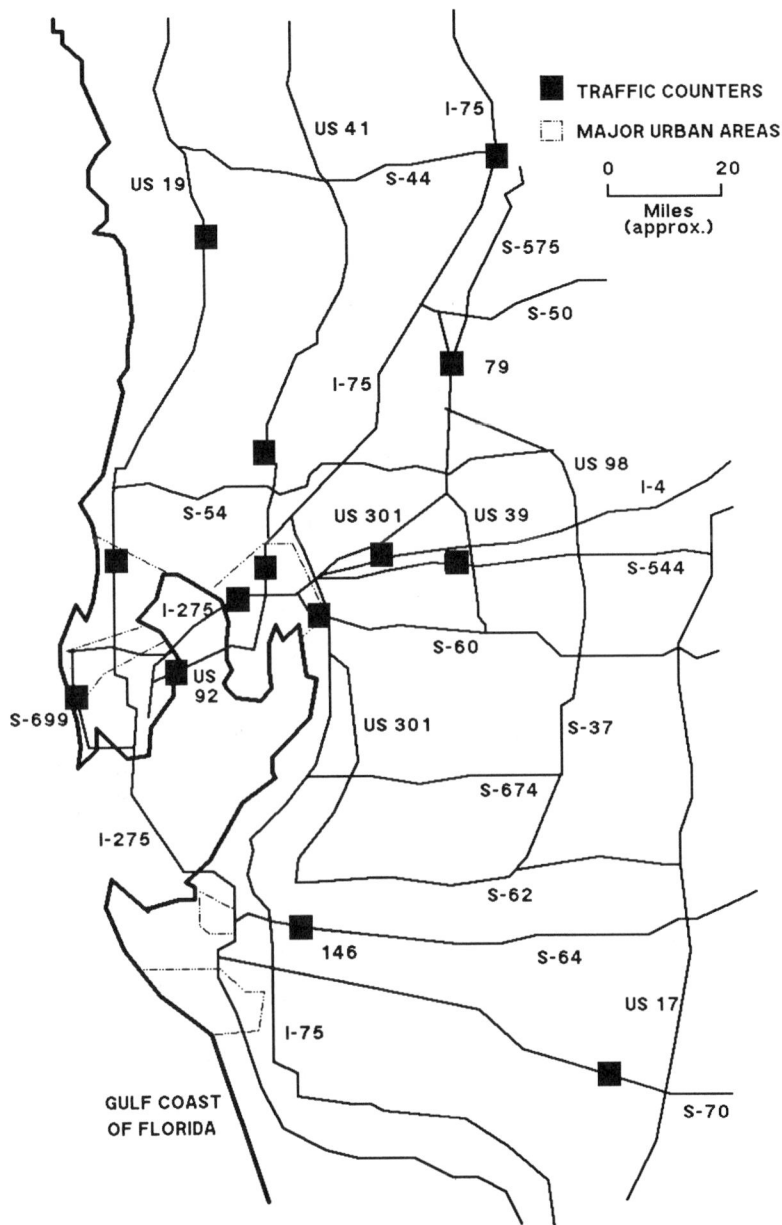

Figure 6.3. Location of traffic counters in Tampa Bay area.

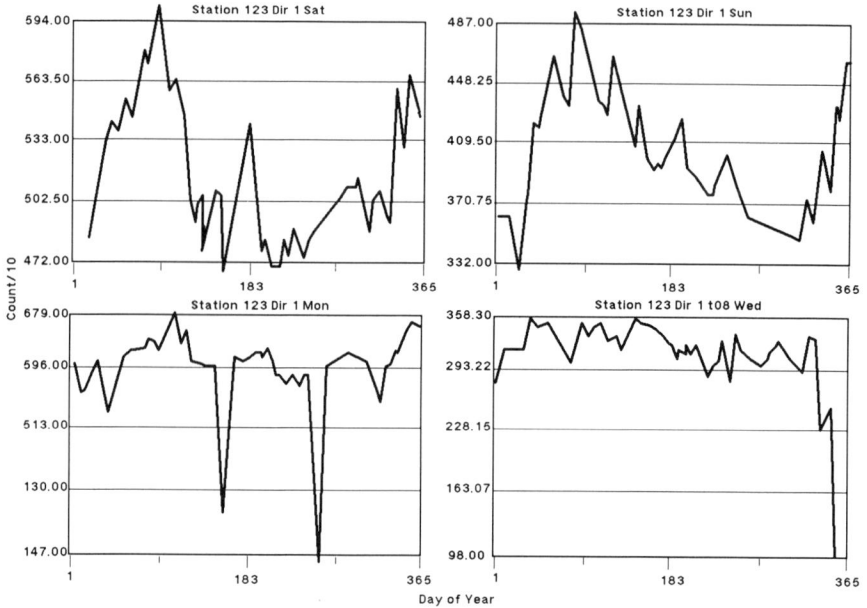

Figure 6.4. Traffic counts during selected days in 1985.

The totals below the table show daily eastbound traffic volumes of 27290 and 25213 respectively, and the "Difference" value of 2077 shows an area "loss" of 2077 more vehicles on August 30 than on the more typical Friday of August 23. (An identical calculation could be applied by RTMAS to up to 24 consecutive days).

In general, daily traffic passing over most two directional counting stations appears to provide approximately zero net gains or losses to the region. This is as expected, except for certain special event days. Once computed, these net differences in traffic flow are then used as the baseline against which to establish net vehicle based evacuation during threat build-up (requiring vehicle utilization data in order to provide an estimate of number of people leaving the urban area: see Southworth, Chin and Cheng, 1987).

Upon AUTOBOX model selection and calibration, which is automatic as far as the RTMAS user is concerned, the user is offered a very flexible set of graphing options, one of which is shown in Figure 6.7. Shown here are the east-bound (inland) traffic counts on Florida's Interstate 4, for those Fridays between July 26 and August 30, 1985. The lower of these two plots is an elongated version of the final, forecast series shown in the top diagram (i.e. for August 30, 1985), showing the observed historical data counts, the model predicted counts, and upper and lower 95% confidence intervals on this model prediction. The resulting univariate time-series model is represented graphically in Figure 6.7.

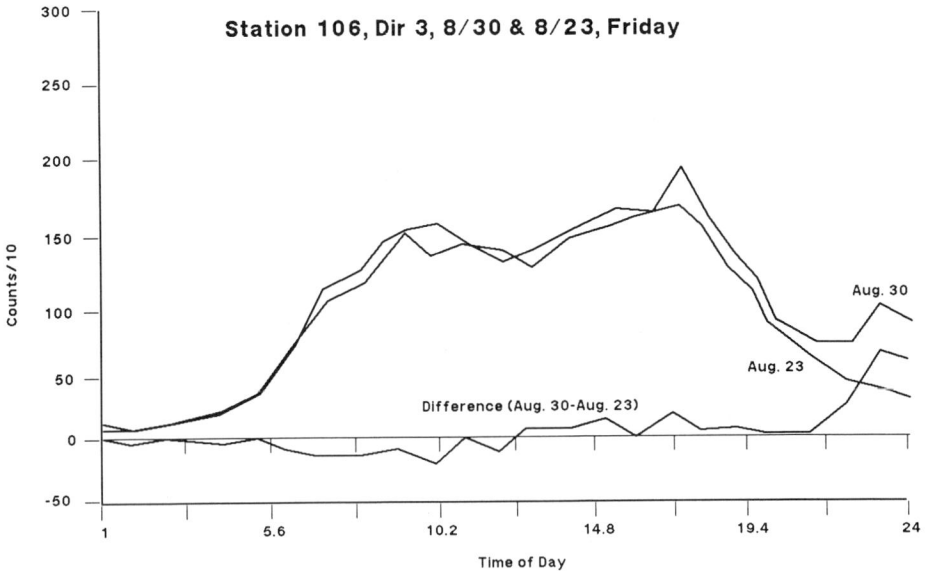

Figure 6.5. Estimated traffic profiles for example analysis.

To handle such multiple events (on multiple highways across the region), the ultimate goal for RTMAS is to become an artificially intelligent rule based expert system, able to offer advice on the speed with which such as evacuation is developing, compensating for other activities going on in the area on the basis of its programmed memory of such events.

Conclusions

Work is currently underway to extend RTMAS to incorporate a fifth software component, a traffic assignment based evacuation simulation model (recall Figure 6.2). This simulation model will be used to predict traffic congestion "hot spots" before they occur, based upon the latest telemetrically received counter data (see Southworth, Chin, and Cheng, 1987). An example application of such an evacuation simulation model is described in Southworth and Chin (1987), where representative multi-color graphics screens showing evacuee route, highway link, and destination selections are reproduced. Given, say a record of the first two hours of an evacuation the above rule based expert system might be extended to activate such a simulation of the region's predicted traffic evacuation pattern (recognizing designated evacuation routes, where appropriate).

Station: 106   Date: 8/23/85

	t01	t02	t03	t04	t05	t06	t07	t08	t09	t10	t11	t12
#3	188	92	124	161	365	751	1206	1432	1590	1424	1486	1319
#3	146	105	123	183	349	811	1319	1529	1629	1659	1486	1415
Dif	42	-13	1	-22	16	-60	-113	-97	-39	-235	0	-96
	t13	t14	t15	t16	t17	t18	t19	t20	t21	t22	t23	t24
#3	1454	1581	1791	1718	2321	1778	1423	1074	848	868	1251	1045
#3	1335	1504	1633	1743	2043	1706	1324	1014	807	548	446	356
Dif	119	77	158	-25	278	72	99	60	41	320	805	689

Total # 3 =27290   Total # 3 =25213   Difference = 2077

Figure 6.6. Sample system screen display showing hourly traffic counts (August 23, 30; Interstate 4, Eastbound).

One purpose of this simulation modeling exercise is to fill in estimates of the traffic volumes on those highways not providing count data, including many secondary arterial and major collectors. With considerable evidence that large scale evacuations tend to generate very few accidents and display much less panic than might be supposed (Federal Emergency Management Agency, 1984), quite accurate simulation results might be expected. Given sufficient data on actual evacuation or some combinations of actual and simulated evacuations, a set of rules may be applied to this data to select the earliest point at which the system might reasonably be expected to provide an accurate prediction of the traffic situation throughout the remaining portion of the evacuation.

Especially useful to emergency management officials is early warning of the number of evacuees (the "spontaneous" portion) already leaving the area, prior to official calls to evacuate. By teaching RTMAS to recognize such spontaneous evacuation and then infer from such behavior the likely traffic pattern or risk related outcomes, it may be possible to operationalize a dynamic traffic rerouting policy, given telephone (possible cellular phone) communication with on the scene traffic police. While appropriate command and control protocols will have to be developed for effective cooperation between the regional agencies involved, the rule based expert system approach proposed here suggests that such real time monitoring and control of emergency evacuations can become reality in the not too distant future.

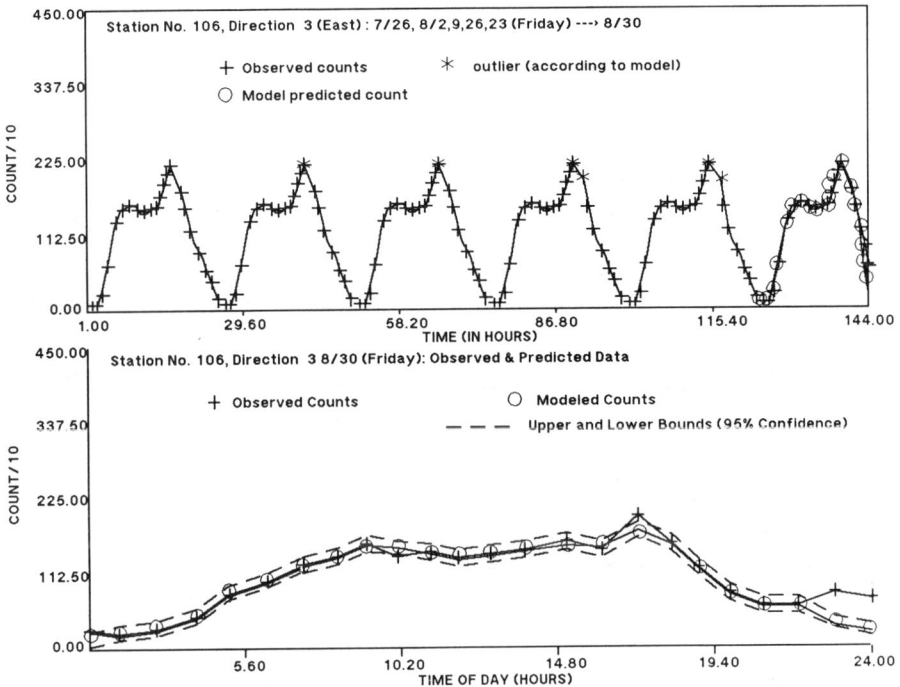

Figure 6.7. Univariate time series model for the example analysis.

Acknowledgements. This research work was sponsored by the Federal Emergency Management Agency, Washington D.C., in cooperation with the Office of Highway Information, Federal Highway Administration, Washington D.C., and the Traffic Counts division, Florida Department of Transportation, Tallahassee, Florida. All opinions expressed and any errors are, however, solely attributable to the authors of the paper. All references in this report to the GURU software package refer to Micro Data Base Systems, Inc.'s GURU Beta testing Version 1.10.06. Micro Data Base Systems Inc., P.O. Box 248, Lafayette, Indiana 47902. All references to the CROSSTALK software package refer to Microstuf's CROSSTALK XVI. Microstuf, 1000 Holcomb Woods Parkway, Suite 440, Roswell, Georgia 30076. All references to the Autobox software package refer to Automatic Forecasting Systems, Inc.'s Autobox version 1.02. AFS, Inc., P.O.Box 536, Hatboro, Pennsylvania 19040.

References

Ahmed, M.S. and A.R. Cook, 1979. "Analysis of freeway traffic time series data using Box and Jenkins techniques," *Transportation Research Record*, 722, 1-9.

Ahmed, M.S. and A.R. Cook, 1982. "Application of time-series analysis techniques to freeway incident detection," *Transportation Research Record*, 841, 19-21.

Bell, W., 1983. "A computer program (TEST) for detecting outliers in time series," in *Proceedings of the American Statistical Association,* Business and Economic Statistics Section, pp 624-639.

Benjamin, J., 1986. "A time-series forecast of average daily traffic volume." *Transportation Research,* 30A, 51-60.

Box, G.E.P. and G.M. Jenkins, 1970, 1976. *Time Series Analysis: Forecasting and Control,* Holden-Day, San Francisco.

Box, G.E.P. and G.C. Tiao, 1975. "Intervention analysis with applications to environmental problems." *Journal of the America Statistical Association,* 70, 70-79.

Chang, I. and G.C. Tiao, 1983. "Estimation of time series parameters in the presence of outliers," *Technical Report No. 8,* University of Chicago, Statistical Research Center.

Davis, G.A. and N.L. Nihan, 1984. "Using time-series designs to estimate changes in freeway level of service, despite missing data," *Transportation Research 18A,* 431-438.

Eldor, M., 1977. "Demand predictors for computerized freeway control systems," *In Proceedings of 7th International Symposium on Transportation and Traffic Theory,* (edited by T. Sasaki and T. Yanoka), 341-270.

Federal Emergency Management Agency, 1984. "Behavior and attitudes under crisis conditions," Selected issues and findings, Washington D.C.

Heathington, K.W., R.D. Worrall, and G.C. Hoff, 1970. "An evaluation of the priorities associated with the provision of traffic information in real time," *Highway Research Record 336,* 15-17.

Helliwell, J., 1986. "Guru: brave new expert system?" *PC Magazine,* 5, No.10, 151-163.

Kalman, R.E., 1960. "A new approach to linear filtering and prediction problems." *Journal of Basic Engineering, ASME* Transactions 82, Part D, 35-45.

Nihan, N.L. and K.O. Holmesland, 1980. "Use of Box and Jenkins time series technique in traffic forecasting," *Transportation 9,* 125-143.

Reilly, D.P., 1984. "Automatic intervention detection system," *Automatic Forecasting Systems, Inc.* P. O. Box 536, Hatboro, Pennsylvania 19040.

Tampa Bay Regional Planning Council, 1984. "Tampa Bay Region Hurricane Evacuation Plan," *Technical Data Report Update,* St. Petersburg, Florida.

Southworth, F. and S-M. Chin, 1986. "Quantifying spontaneous evacuation in time of threat: a feasibility study," Prepared for the Federal Emergency Management Agency, Washington D.C.

Southworth, F. and S-M. Chin, 1987. "Network evacuation modeling for flooding as a result of dam failure," *Environment and Planning A,* 1543-1556.

Southworth, F., S-M. Chin, and P.D. Cheng, 1987. "Quantifying spontaneous population evacuations in time of threat: a real time traffic monitoring system for the Tampa Bay area," Prepared for the Federal Emergency Management Agency, Washington D.C.

Tiao, G., 1986. "ARIMA models, intervention problems," *Technical Report No. 27,* University of Chicago, Statistical Research Center.

Tsay, R.S., 1986. "Time series model specification in the presence of outliers," *Journal of the American Statistical Association,* 81, No. 393, pp 132-140.

Expert Systems in Site Selection

Introductory remarks by Lyna L. Wiggins

In the first chapter in this book, Ortolano and Perman devote a lengthy section to the importance of selecting the right task for an expert system. Important criteria discussed in that section include that: (1) the knowledge needed is specialized and narrowly focused; (2) true "experts" exist; (3) the task is neither trivial nor exceedingly complex; (4) conventional (algorithmic) programs are inadequate; (5) the potential payoff is significant; and (6) an articulate expert is available and willing to participate. In Part Three of this book, we have included chapters that describe prototype expert systems designed to address the task of site selection. Site selection tasks are also addressed, from slightly different perspectives, in Chapter 3 (Wright) and in Chapter 13 (Lee and Wiggins). It is interesting that researchers, all working independently, but each concerned with the application of expert systems to urban planning, have selected this particular task as one that is appropriate to expert system implementation. The task does seem to have most of the 6 characteristics listed above. The reader will see that each of the authors in the next 4 chapters selects slightly different aspects of the site selection problem on which to focus in their system development.

In Chapter 7, Findikaki describes a system she developed as an aid to decision makers faced with a site selection problem. The Site Selection Expert System (SISES) has been designed to aid land use planners, developers, or prospective land users. The system helps a user develop a set of site attributes and determine their weighted relative importance. The knowledge base of SISES consists of 2 parts: (1) knowledge representing basic facts about site selection; and (2) heuristic knowledge induced by observing specific examples of judgment made by the expert decision maker. The latter information is obtained by having the user rank order a set of site profiles based on the selected attributes. Then a variation of conjoint analysis, using paired comparisons, is used to derive the user's importance weights for the attributes. The system is written in Pascal and implemented on a microcomputer.

In Chapter 8, Suh, M.P. Kim and T.J. Kim describe a prototype expert system for site selection of manufacturing establishments funded by foreign investors. The Expert System for Manufacturing Site Selection (ESMAN) makes use of existing survey data collected from managers of existing foreign manufacturing facilities in the United States. The survey data from these experts identifies location factors and measures the respondents' importance ratings of these factors. These importance ratings are combined with information about the specific facility under consideration, obtained during a session from the user (e.g., type of products, number of employees, size, etc.). The prototype system contains a relatively large data base including 19 tables of factor weights, and a site characteristic table including 34 characteristics for 3 potential sites. The prototype demonstrates that not only can the knowledge and expertise of an urban and regional planner be codified in a system that can advise less experienced planners, but also that such a system can become an educational tool that benefits urban and regional planning students. The system was developed using the expert system shell Personal Consultant Plus and implemented on a microcomputer.

In Chapter 9, Han and Kim explain the concept and development of another site selection and analysis system. The Expert System for Site Analysis and Selection (ESSAS) is a prototype system with 240 decision rules, and is designed to solve a portion of the site selection problem. The system was developed to aid master planners for the Army who are responsible for planning future building projects at military bases. Site selection models based on optimization approaches have been largely unsuccessful since the optimal land use plans developed by these models often do not fulfill other important criteria such as regulations, guidelines and other judgmental concerns. Similarly, there are methodological difficulties in traditional siting procedures which rely on McHarg-type overlay procedures. To incorporate intuitive judgments into the site selection process, an expert system approach seems appropriate. Information from regulations, field manuals, design guides and interviews with expert master planners was combined into the final ESSAS system. Currently the system is designed to aid only in the siting of administrative office buildings. The success of the prototype suggests that the overall approach is viable and that a more complete system, including a variety of land uses, is achievable. The system was developed using the expert system shell Personal Consultant Plus and implemented on a microcomputer. Graphics, including maps generated using AutoCad, are incorporated into the system.

In Chapter 10, Rouhani and Kangari describe the development of an expert system for landfill site selection. The system is intended to be used as a consultant to a waste manager/planner during the preliminary phases of site selection. The process of designating a location as a landfill requires extensive studies of biogeochemical and hydrologic characteristics of all potential sites. This process demands a considerable amount of empirical input, expert opinion and heuristic rules. The authors present a prototype expert system for site selection with rules based on U.S. Environmental Protection Agency documents for the ranking of uncontrolled hazardous waste sites for remedial actions. The system includes rules concerning ground water routes, local climate, waste characteristics, planned features of the proposed landfill facility and targets at risk.

The system user must select appropriate answers to such questions as the depth of the aquifer, local net precipitation, projected waste quantity, etc. The final output is a relative ranking of a set of candidate sites. The system was developed using the expert system shell Insight 2+ and implemented on a microcomputer.

CHAPTER 7

SISES: An Expert System for Site Selection[1]

Irene Findikaki

Site selection is a critical decision making element for a wide range of economic activities, from land use planning to residential location selection by individuals. Typically, site selection involves two phases. First, a small number of alternatives among all possible sites are identified for further evaluation. Then these selected alternatives are examined in depth to find the optimal site. Because of the effort required for a thorough evaluation of each site, the number of sites included in the second stage is, by necessity, small. Therefore, the first phase of the site selection process is very important because if, for example, the "best" available site is not included in the set of alternatives selected for thorough study, then this site can never be selected as the optimal site. In large scale planning the first phase involves reconnaissance and prefeasibility studies. At the individual level, it may involve consultation with a specialist.

This paper presents a methodology for systematizing and expediting the first of the above two phases in the site selection process. The proposed methodology is implemented in the expert system SISES (Site Selection Expert System) that can be used as a decision making tool in selecting a site for a specific use. SISES is designed to capture the expertise of expert decision makers without making any prior assumptions about their preferences, background, or identity. For example, the decision maker can be a land use planner, a developer, or a prospective land user. Although the system was developed to reproduce a specific domain of expertise, it can be easily adapted to other applications with a similar conceptual framework independently of domain. The system has the

1   The original version of this chapter was published in *Experts Systems in Civil Engineering*, Proceedings of a Symposium of the American Society of Civil Engineers, M.L. Maher and C.N. Kostem (eds), April 8-9, 1986, pp. 182-192.

ability to analyze and learn about the decision making process of individuals.

SISES is written in Pascal and runs on a microcomputer. The present paper gives an overview of the ongoing development of SISES, and focuses on the knowledge acquisition methodology.

The Structure of SISES

SISES consists of four parts: the knowledge acquisition, induction, design, and decision analysis units. The knowledge acquisition unit is used to collect and organize information provided by expert decision makers. The induction unit evaluates this information and generates rules and entity evaluation functions expressing the expert judgement. Knowledge acquisition in SISES is based to a great extent on a multiple trade-off data gathering procedure and requires a close interaction between the knowledge acquisition and the induction units. The design unit can customize the knowledge of the acquisition system to particular applications by modifying selected elements of the knowledge acquisition unit and expanding it with the addition of new components. The decision unit uses the rules generated by the induction module and employs decision theory techniques for selecting one or more of the available alternatives. The overall structure of SISES is illustrated in Figure 7.1.

The knowledge acquired from an expert can be stored for future use, thus contributing to a user's incremental learning of a subject. The system offers the option of combining knowledge from several experts and storing it in a dynamic knowledge base which grows as new decision makers are interviewed. The system can use either the selection rules of a single decision maker, or rules representing the idiosyncratic decision processes of a group of individual decision makers. Therefore, points of departure among the views of human experts who disagree can also be represented and taken into account.

The Knowledge Acquisition Methodology

Knowledge acquisition is the most important of the central problems in artificial intelligence research (Fiegenbaum, 1983). Quinlan (1979) points out that one of the reasons that knowledge acquisition is the bottleneck of expert system building is the fact that typically experts are asked to perform tasks that they do not ordinarily do, such as describing everything they know about their domain of expertise. A key research objective in knowledge acquisition is the development of methods of machine learning to replace the conventional approach of knowledge acquisition via dialogue, in which the expert authors rules during long interviews with a knowledge engineer. Machine learning involves the development of rules by analyzing cases of correct decision making, and generalizing from specific examples provided by the expert. Induction of rules by

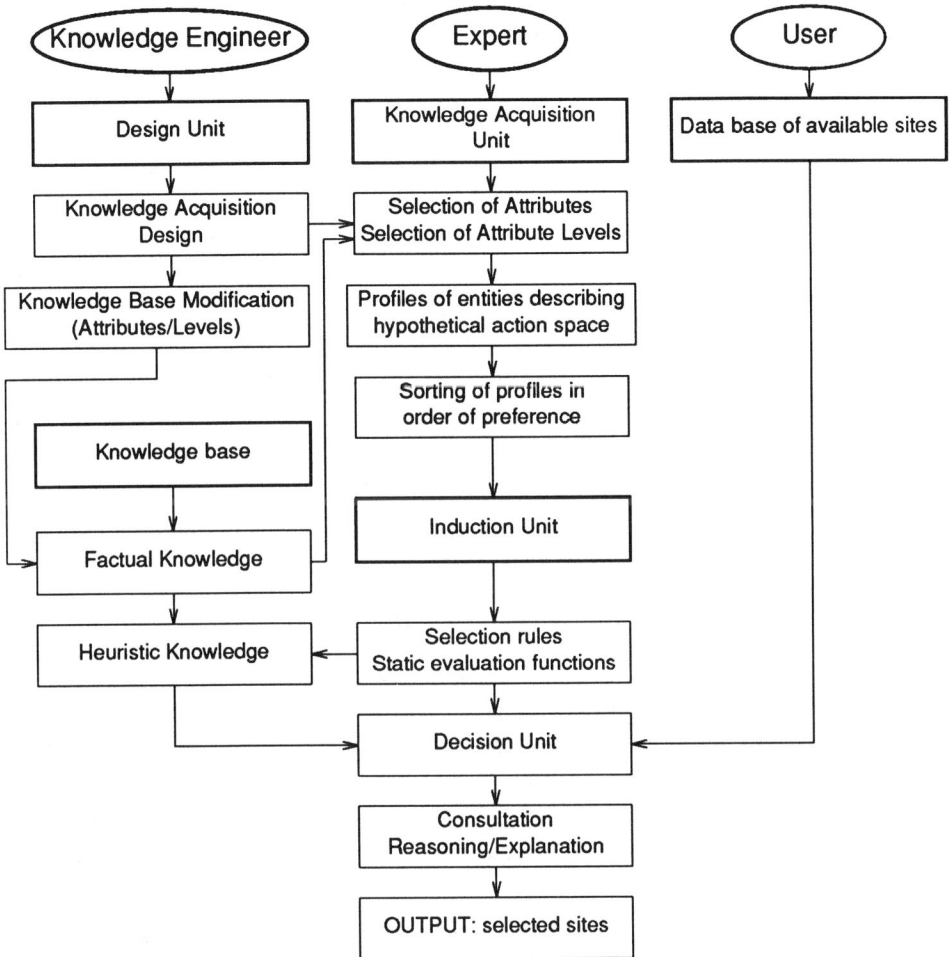

Figure 7.1. Schematic Structure of SISES.

generalization from examples of expert decision making has been used success-fully in some expert systems (Mitchie et al., 1984).

The knowledge base of SISES consists of knowledge representing basic facts about site selection built in the system, and heuristic knowledge which is induced by observing and analyzing specific examples of judgmental behavior of the expert. The knowledge acquisition unit is designed to interact with the expert to acquire information relevant to both the knowledge domain and the judgmental behavior of the expert.

The knowledge acquisition unit is complimented by a design unit which can modify the default data base of factual knowledge used by the system, update information about the domain, and alter the design of the knowledge acquisition unit.

The knowledge acquisition methodology is based on the following assumptions on the behavior of an expert decision maker:

a. An expert's judgement on the selection or rejection of a site for a particular use is determined by a limited number of basic factors or attributes which are a subset of all site characteristics. These factors form the profile of a site.

b. An expert's judgement is based on the evaluation of several alternative sites described by the levels of their attributes. The decision process leading to the selection of a particular site is governed by simultaneous trade-offs between pairs of attributes and between entire site profiles. The final judgement of the expert is the product of the evaluation of complete profiles of the available sites.

c. Site selection is made under conditions of certainty and is of extreme importance to the decision maker. Consequently, the individual decision maker always tries to select the best available alternative, i.e. the alternative perceived as having the maximum value. This eliminates the need for probabilistic modeling of the expert's judgement (Shoker and Srinivasan, 1979; Johnson 1974). This assumption implies zero probability of selecting any alternative other than the one representing the expert's top choice. A consequence of this assumption is that preferences can be measured on an ordinal scale.

d. The perceived value of a site is a function of the characteristics of the site weighted by their relative importance to the decision maker.

e. The order of preference among multi-attribute entities represents information about an individual's decision process in condensed form. Decomposition of the overall judgement of an individual into its elements provides the ground for making inferences about the relative importance of each attribute and for analyzing the trade-offs which characterize the decision process.

f. Although direct statements by decision makers about preferences, rules of thumb, and comparison procedures are valuable, in most cases it is uncertain whether the decision makers are fully aware of the facts, prejudices, beliefs and heuristic knowledge influencing their judgement. Declarative statements do not always provide the information needed to reproduce the decision maker's behavior.

The description of a decision maker's judgmental process is achieved in SISES by combining information based on declarative statements and information induced from observing the decision maker's behavior in specific examples. The knowledge acquisition unit is designed to provide the system with both types of information. The latter type of information is obtained by a special method which reveals the preference structure of individual decision makers. Some of the above cognitive process assumptions are incorporated in a function which expresses the outcome of the judgmental evaluation of individual entities within a hypothetical action space. This function, which quantifies the value of individual entities to the decision maker, is termed the entity evaluation function.

The hypothetical action space is an abstraction of a subset of the real action space of the decision maker. The real action space is defined as the part of the environment that the individual has contact with (Wolpert, 1965). The concept of the action space is similar to the concept of "life space" introduced by Lewin (1951). The hypothetical action space consists of a finite number of entities, typically much smaller than in real action space, described by combinations of discrete levels of attributes that represent the decision maker's real action space. The system generated hypothetical test space is designed to include entities covering the entire range of entity attributes found in the decision maker's real action space. For example, in the case of residential selection, a hypothetical action space may consist of several profiles of residential sites, each described by attributes that are important to the decision maker, e.g. price, air quality, noise level, quality of schools, density, proximity to main arteries, distance from work, etc. (Findikaki, 1981, 1982). SISES uses an experimental design scheme to reduce the dimensionality of the hypothetical action space to a manageable number of dimensions. The default experimental design is a factorial design (Addelman, 1962a,b).

The system's level of the knowledge acquisition session is selected depending on the desired level of sophistication, time limitation, and availability of the expert. In developing a sophisticated system and working with an expert whose availability is not a limiting factor, a session based on the full-fledged knowledge acquisition design should be selected. Alternatively, a shorter session based on a less elaborate design could be used for a less sophisticated system.

The first step in a typical knowledge acquisition session is the selection of the attributes that describe the entities forming the hypothetical action space. For this purpose, the expert is asked first to specify the type of land use in the problem. Then the most important site attributes for the desired land use are identified through a series of questions to the decision maker. The list of potential attributes for different land use types is part of the factual knowledge base of the system. This list can be expanded with additional attributes suggested by the expert. New attributes proposed by the expert during a particular session can be used to expand the permanent data base of potential attributes. After the relevant site attributes are identified, the expert is asked to distinguish between quantifiable attributes and criteria.

Next, the system asks the expert to verify the range of potential values of the quantifiable factors. The default values for the range of the attributes provided by the system are displayed, and the expert is asked to define the lowest and the highest expected level of each new attribute. The expert also may change any default attribute level ranges.

Then, the expert is asked to group the quantifiable attributes in groups of five or six, and rank the groups in order of importance. For example, in the case of sixteen total attributes, the expert would identify three groups of attributes, the six most important, the five next most important, and the five least important attributes. The attributes in each group are then used by the system to construct profiles of individual entities forming a hypothetical action space. Each profile is described by the levels of the attributes. The level of each attribute is determined with the aid of a fractional factorial design.

Then, the expert is presented with the set of site profiles defining the hypothetical action space formed by the first group of attributes. These profiles are ranked by the expert in order of preference. At this point, the system passes the ranking of the profiles to the induction entity which, using the method described in the next section, then identifies the most important among the attributes in this group. The most important attribute is used together with the next group of attributes to design a second set of hypothetical profiles which the expert is again asked to rank. Then the most important attribute of the first group is used with the remaining attributes to form a third set of hypothetical profiles and the expert is asked again to rank them.

A similar approach is followed with the non-quantifiable attributes. The effect of these attributes is expressed in the form of yes/no conditions. The expert is asked first to identify the non-quantifiable attributes and the corresponding yes/no condition that would result in the rejection of a site.

The Induction Unit

The effect of the non-quantifiable attributes on the expert's judgement can be expressed by a set of rules. These rules can either be based on declarative statements, or induced from specific examples of expert decision making.

The combined effect of the quantifiable attributes is described by a linear function of weighted attribute levels. This formulation is based on the assumption that the cognitive process of an individual making choices can be simulated by a linear model (Fishbein, 1954; Rosenberg, 1976). Thus, the value of a particular site to the decision maker is expressed as:

$$\text{(Site Value)}_j = \sum_{p=1}^{n} W_p X_{jp},$$

where n is the number of site attributes, X_{jp} is the level of attribute p for site j, and W_p is the importance of attribute p. The value of a site may increase as the level of each attribute p increases (if $W_p > 0$), or as the level of attribute p decreases (if $W_p < 0$), or it may be independent of the level of p (if $W_p = 0$). The selection criterion is that the expert selects the site which has the largest value of the evaluation function.

The order of the profiles constitutes a set of conjoint measurements reflecting the joint effect of different attributes on judgments (Tversky, 1967a,b). The order of preference of the profiles is converted to paired judgement comparisons which are analyzed to obtain the importance weights of the attributes forming the entity evaluation function of the individual. The importance weights are estimated to reproduce the order given by the expert. This is achieved by using a linear programming formulation (Shoker and Srinivasan, 1977) in which the objective is to minimize the error in reproducing the expert's ordering of the profiles and the constraints are based on the paired judgement comparison.

This method is applied to all sets of ordered profiles. Then, the most important attribute which is included in the description of all hypothetical spaces is used as a bridging factor to combine the evaluation function for each action space into a global evaluation function.

The Decision Making Unit

The decision making unit evokes different scripts matching the user's needs. The system composes these scripts using a combination of built in information, induced rules, and entity evaluation functions.

During consultation, the user is asked to provide a description of all potential sites in terms of the attributes used by the system for the particular land use under consideration. The system first makes inferences to reduce the data base of potential sites by eliminating those that should be rejected according to the induced rules based on the non-quantifiable site attributes. Then the system computes the entity evaluation function for each of the remaining sites. The output of the consultation sessions consists of the order of preference of the selected sites and an optional explanation for rejecting the rest.

Summary and Conclusions

SISES is an expert system for site selection that can be used as a decision tool in land use planning, residential selection, and other site dependent planning. The system consists of four units for knowledge acquisition: rule and evaluation function, induction, decision making, and design of customized versions of the knowledge acquisition unit. The emphasis of the present work is on the description of the knowledge acquisition unit. The knowledge acquisition methodology used by SISES is based on the interactive generation of a set of profiles describing a hypothetical action space that the expert is asked to rank. The ranking of these profiles is decomposed to paired comparison judgments which constitute a set of examples of expert decision making. Analysis of these examples leads to the induction of site selection rules and the generation of an entity evaluation function. The selection rules and the entity evaluation function are applied on the data base with the characteristics of all available sites provided by the user to generate the list of selected sites, ranked in order of preference.

References

Addelman, S., 1962a. "Orthogonal Main-Effect Plans for Asymmetrical Factorial Experiments", *Technometrics*, Vol. 4, pp. 21-26.

Addelman, S., 1962b. "Symmetrical and Asmmetrical Fractional Factorial Plans", *Technometrics*, Vol. 4, pp. 47-59.

Fiegenbaum, E.A., 1983. "Knowledge Engineering: The Applied Side", in *Intelligent Systems*, J.E. Hayes and D. Michie, eds., Ellis Horwood Limited, Chichester, England.

Findikaki, I.T., 1980. "Residential Location Preferences: A Conjoint Analysis Approach," *Engineer Thesis*, Stanford University.

Findikaki, I.T., 1982. "Conjoint Analysis of Residential Preferences," *PhD. Dissertation*, Stanford University.

Fishbein, M., 1967. "A Behavior Theory Approach to the Relations between Beliefs about an Object and the Attitude towards the Object", in *Readings in Attitude and Theory Measurement*, M. Fishbein ed., John Wiley and Sons, New York.

Johnson, Richard M., 1974. "Trade Off Analysis of Consumer Values" in *Journal of Marketing Research*, Vol. XI, May, 1974.

Lewin, K., 1951. *Field Theory in Social Sciences*, Harper and Row, New York.

Michie, D., S. Muggleton, C. Riese and S. Zurbrick, 1984. "Rulemaster: A Second Generation Knowledge-Engineering Facility", *First Conference of Artificial Intelligence Applications*, Denver, December 5-7.

Quinlan, J.R., 1979. "Discovering Rule by Induction from Large Collection of Examples" in *Expert Systems in the Micro Electronic Age*, D. Mitchie ed., Edinburgh University Press.

Rosenberg, M.J., 1956. "Cognitive Structure and Attidunal Affect", *Journal of Mathematical Psychology*, Vol. 4, pp. 1-20.

Tversky, A., 1976. "A General Theory of Polynomial Conjoint Measurements", *Journal of Mathematical Psychology*, Vol. 4, pp 175- 201.

Wolpert, J., 1965. "Behavioral Aspects of the Decision to Migrate", in *Papers of the Regional Science Association*, Vol. 15, pp. 159-169.

CHAPTER 8

ESMAN: An Expert System for Manufacturing Site Selection[1]

Sunduck Suh, Moonja Park Kim
and Tschangho John Kim

Since the establishment of the Office of Foreign Investment in 1975 in the United States Department of Commerce, foreign inward direct investments in the U.S. have rapidly increased. Foreign inward direct investments include those of nonbank U.S. affiliates of foreign companies. A U.S. affiliate is a U.S. business enterprise in which there is foreign direct investment; that is, in which a single foreign person owns or controls, directly or indirectly, ten percent or more of the voting securities of an incorporated business enterprise or an equivalent interest in an unincorporated business enterprise (U.S. Department of Commerce, 1985). In 1984, non-Americans had 527.6 billion dollars worth of direct investment in the U.S., of which about one-fourth was used in some form of manufacturing activity. Total foreign direct investment in the U.S. in 1974 was 26.5 billion dollars.

Foreign inward investments in the U.S. have been welcomed because

1. they create jobs;

2. they induce transfer of foreign technology from abroad to the U.S.;

3. they have a positive impact on the U.S. balance of payments;

4. they broaden the tax bases at all levels; and

5. they help to stabilize the U.S. dollar (U.S. Congress, 1977).

---

1 The original version of this chapter was published in *Computers, Environment and Urban Systems*, 1988, 12(4):239-252.

Table 8.1 shows a partial list of Japanese inward direct investments in the U.S. in the 1980s.

Table 8.1. Partial List of Japanese Establishments

Company	Produce	Asset/Capacity	Location	Year Open
Honda	Automobile	2423 Workers 875 car/day	Maryville, Ohio	1983
Fujitsu America	Magnetic Disk	$ 30 million	Portland	1986 (under construction)
Epson	Printer Assembly Plant	$ 10 million	Portland	1986 (under construction)
Sharp & RCA	Semi-conductor	$260 million	Camas, Washington	
Mazda	Automobile	$450 million	Flat Rock, Michigan	
Mitsubishi Motor	Automobile		Bloomington Illinois	1985 (under construction)
NEC	Telecommn Equipment		Portland	
Hitachi	Semiconductor		Dallas	
Toyota	Automobile	$800 million	Kentucky	mid-1988

Source: Forbes, 1985, 1986

  One of the most crucial decisions that industrial management has to make in considering establishment of a new plant is plant site selection. How does a foreign inward investor make a decision on plant location? Tong (1978) surveyed managers of existing foreign manufacturing plants in the U.S. in order to analyze the comparative weight of thirty-two location factors. The list of location factors as well as the importance of these factors (expressed in terms of a mean rating on a 1 to 5 scale) for all responding firms is shown in Table 8.2. The main outcome of the survey is a series of tables that shows mean rating of location factors by national origin, by each of twelve different industries, by thirteen U.S. states, by three employee-size categories, and by four degrees of foreign ownerships. For example, availability of transportation service was ranked highest for all responding firms for Canadian, Swedish, and Swiss manufacturing investors, but ranked as the second for French investors, fourth for Japanese investors, and eighth for German investors. For English manufacturing investors, labor attitude ranked first. Table 8.3 shows the ranking on five factors by each of thirteen different industries. Importance ratings found by Tong (1978) became a basis for designing a prototype expert system for the ESMAN which is described in the forthcoming section in detail.

Building an Expert System for Site Selection

The need for the development of the prototype expert system, "Expert System for Manufacturing Site Selection" (ESMAN) comes from state or local planners who must identify possible sites when asked by potential foreign manufacturing investors. Frequently, state and local officials have been asked to identify suitable sites and provide appropriate information to foreign manufacturing investors, who are looking for suitable locations for production plants.

A prototype expert system of ESMAN was developed using Personal Consultant Plus Version 2.0 running on a Compaq 286 portable computer. The schematic diagram of the system is summarized in Figure 8.1. In a typical consultation, a user supplies plant-related data, such as kind of products, number of employees, investment size, transportation system, market information, utility requirement, and other data. The first step the system considers is if a potential site has a lot large enough for the user's requirement. If a potential site does not have an appropriate lot available, this site is excluded from further analysis and the system considers the next potential site. The system continues the further analysis only for the potential sites that contain the size lot required.

The second step the system performs is selecting appropriate weights data from the reference data base depending on the initial user's input. The reference data base was selected from a few hundred tables in Tong (1978) as described before. The thirteen components were derived from thirty-two location factors used in Tong (1978). For thirteen components, weight tables for Product Type (Table 8.4), Foreign Ownership (Table 8.5), and Employment Size (Table 8.6) were used in this study. For example, if a user entered "Textile" as product type, a high percent of foreign ownership, and medium employment size, then the system would select the appropriate column from each table for the weights of the thirteen components.

The third step is to develop an index for each of the thirteen components by applying various rules to the profiles of potential sites. The profile data used in this study is a small sample of three potential sites (Table 8.7).

The fourth step is to calculate a combined weighted index for a site, i.e., Total Index for Site. It was assumed that the combined effect of all the components can be described by a linear function of the weighted individual indices of each component.

Table 8.2. Importance of plant location factors from all responding firms

Factor	Mean Rating	Rank
Availability of Transportation Services	3.701	1
Labor attitudes	3.665	2
Ample space for future expansion	3.652	3
Nearness to markets within the U.S.	3.647	4
Availability of suitable plant sites	3.594	5
Availability of utilities	3.375	6
Cost of suitable land	3.330	7
Attitudes of local citizens	3.290	8
Cost of construction	3.277	9
Labor laws	3.232	10
Availability of skilled labor	3.192	11
Cost of utilities	3.125	12
Salary and wage rates	3.120	13
Cost of transportation services	3.103	14
State tax rates	3.036	15
Local tax rates	2.982	16
Education facilities	2.938	17
Attitudes of government officials	2.924	18
Availability of managerial and technical personnel	2.911	19
Police and fire protection	2.880	20
Availability of unskilled labor	2.781	21
Housing facilities	2.750	22
Proximity to suppliers	2.741	23
Facilities for importing and exporting	2.656	24
Proximity to raw materials sources	2.531	25
Climate	2.375	26
Government incentives	2.321	27
Cost of local capital	2.281	28
Availability of local capital	2.228	29
Nearness to home operation	1.795	30
Proximity to export markets (outside the U.S.)	1.755	31
Nearness to operations in a "third country"	1.219	32

Source: Tong (1978)

Table 8.3. Ranking of selected location components for 13 SIC industries

SIC Component	Transp'n	Labor Attitudes	Future Space	Nearness to Markets	Suitable Sites
Food & Kindred	2	5	3	5	1
Textile	4	3	8	1	6
Paper	3	4	7	1	2
Chemicals	3	5	1	8	1
Rubber	1	2	9	4	5
Glass	1	11	1	16	6
Primary Metals	1	4	8	10	4
Fabricated Metals	7	1	2	3	4
Industrial Machinery	2	7	3	1	5
Electrical Machinery	4	2	3	1	17
Measuring Cont. Equip.	2	1	4	10	10
Other	1	2	3	4	5

Source: Tong (1978)
Note: The same number denotes that those components are tied in rank.

$$\text{Total Index for a Site} = \sum_{j}^{n} W_j I_j$$

where:

n = number of components
I = index of the individual component
W = weights of the individual component.

A sample rule for the above formula is shown below:

```
IF  1) there is a vacant lot big enough for the plant site 1, and
    2) index for local tax rate of site 1 is known, and
    3) index for utilities of site 1 is known, and
    4) index for transportation potential for site 1 is known, and
    5) index for labor condition is known,

THEN    Total index for site 1
        = (index for tax rate of site 1 x weight for tax)
        + (index for utility of site 1 x weight for utility)
        + (index for transportation system x weight for transport system)
        + (index for skilled labor x weight for skilled labor)
        + (index for unskilled labor x weight for unskilled labor)
```

The fifth and final step is to compare the indices of different sites and to recommend alternate sites in rank order. The third step may need to be explained in more detail. For this step, some examples of rules may be helpful

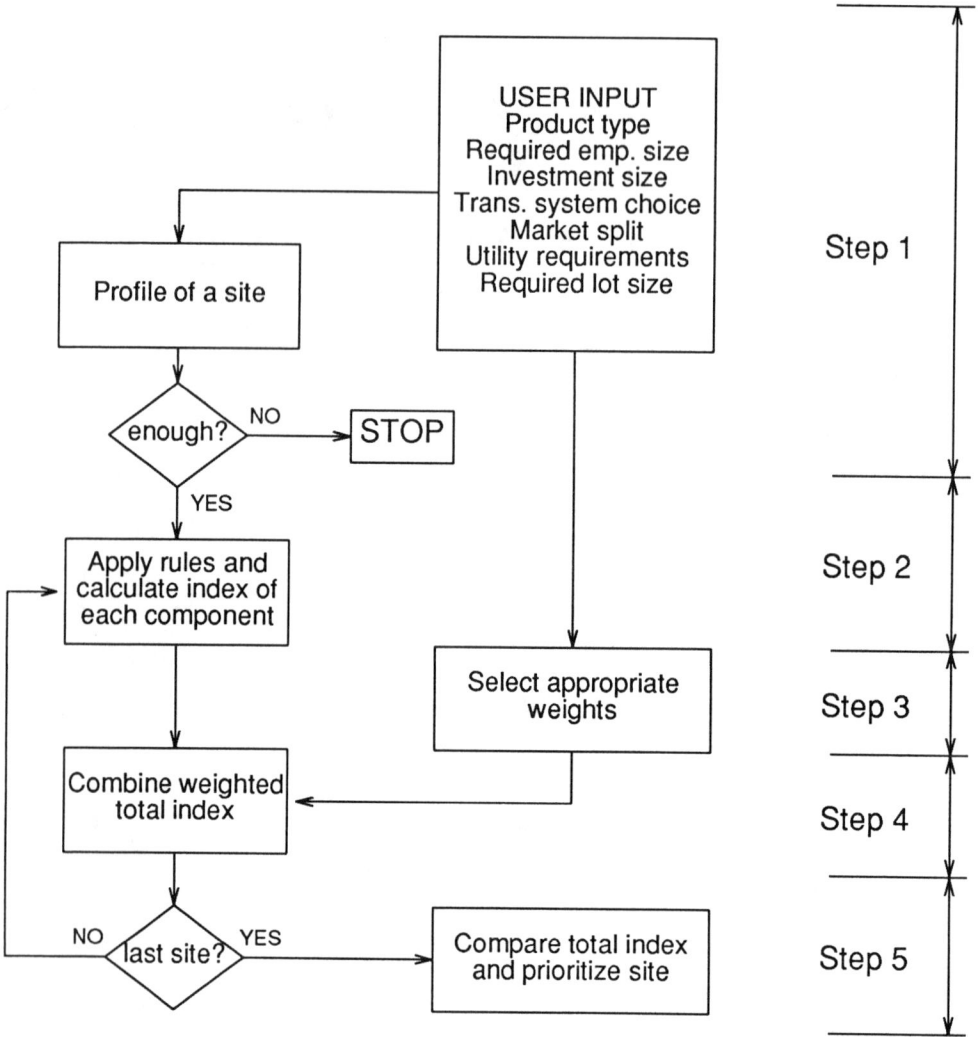

Figure 8.1. Schematic diagram of the ESMAN expert system.

in describing the development of an index for each component. For the "Transport" component, four kinds of rules are used and indices ranging from -100 to +100 were generated for each component by applying the rules. The "Transport" component has three subcomponents, i.e., highway, railway, and air. For the highway subcomponent:

Table 8.4. Weights for 13 Location Components by 12 Product Types.

locational component	Food	Textiles	Paper	Chemicals	Rubber/ Plastic	Glass/ Concrete	Primary Metals	Fabricated Metal	Industrial Machinery	Electrical Machinery	Measuring, Control Equipment	Other
suitable site	4.15	3.35	3.60	4.00	3.36	3.85	3.66	3.52	3.43	3.00	3.20	3.59
skilled labor	2.61	3.07	2.70	3.09	2.90	2.28	3.83	3.42	3.40	3.22	3.40	3.19
unskilled labor	2.61	3.00	3.40	2.50	2.54	3.28	2.33	3.11	2.40	2.72	3.20	2.78
wage	2.84	3.00	2.70	2.84	3.00	3.85	3.16	3.19	3.16	3.22	3.20	3.12
transport	4.00	3.42	3.50	3.87	3.81	4.14	4.33	3.44	3.76	3.38	3.60	3.70
market	3.53	3.85	4.00	3.50	3.45	3.14	3.16	3.66	4.10	3.61	3.20	3.64
utility	3.30	3.21	2.90	3.84	3.36	4.00	4.00	3.36	3.16	3.11	3.20	3.37
tax	3.07	3.14	2.40	2.87	2.90	3.57	2.50	3.00	2.93	3.22	3.20	2.98
capital	2.07	2.35	1.90	1.65	1.81	2.71	2.50	2.63	2.44	3.00	2.22	
police	3.07	2.85	3.10	2.96	2.54	2.71	2.33	2.69	2.90	3.05	3.00	2.88
house	3.00	2.85	2.70	2.75	2.09	2.57	2.16	2.75	2.86	2.94	3.20	2.75
education	3.00	3.28	2.80	3.18	2.27	2.85	2.16	2.83	2.96	3.00	3.40	2.93
climate	2.38	3.00	1.60	2.40	2.36	2.28	2.00	2.11	2.43	2.33	2.80	2.37

Source: Tong (1978)

Table 8.5. Weights for 13 Location Components by 4 Foreign Ownerships.

| Location component | Foreign Ownership | | | |
	low	lower -medium	upper -medium	high
suitable plant site	4.00	4.00	3.63	3.56
skilled labor	2.28	3.75	3.00	3.21
unskilled labor	3.00	3.62	2.96	2.72
salary and wage rate	3.85	4.00	2.92	3.08
transportation service	4.28	4.00	3.63	3.68
nearness to market	3.57	3.50	3.63	3.68
utilities	4.28	3.25	3.70	3.31
local tax rates	3.71	2.75	2.85	2.98
local capital	2.85	2.37	2.18	2.19
police and fire	3.14	2.87	2.85	2.87
housing facilities	3.28	3.12	2.70	2.72
education facilities	3.14	3.37	2.81	2.94
climate	3.14	2.25	2.09	2.40

Source: Tong (1978)

```
IF 1) number of federal highway system serving site > 1,  and
   2) number of state highway system serving the site >1, and
   3) commercial bus service is available, and
   4) package delivery service is available, and
   5) number of freight carriers for interstate movement >0, and
   6) number of freight carriers for local movement >0, and
   7) distance to the nearest interchange = 0,

THEN    index for highway system = 100.
```

For the railway subcomponent,

```
IF      distance to the nearest piggyback facilities = 0

THEN    index of railway system = 100.
```

and finally, for the air system subcomponent,

```
IF 1) length of runway = 8000 feet, and
   2) distance to the nearest airport = 5

THEN    index of air system = 100.
```

These three indices are combined to generate the Transport-Index as follows:

```
IF      transportation mode choice is known,
THEN    index for transportation related condition = (index for
        highway system highway) + (index for railway system
        railway) + (index for air system air system).
```

Table 8.6. Weights for 13 Location Components by 3 Employment Sizes.

Location component	Employment Size		
	small	medium	large
suitable plant site	3.53	3.75	3.78
skilled labor	3.07	3.23	3.78
unskilled labor	2.43	3.31	3.21
salary and wage rate	3.03	3.26	3.39
transportation service	3.67	3.83	3.71
nearness to market	3.79	3.58	3.32
utilities	3.23	3.64	3.53
local tax rates	2.95	2.95	2.28
local capital	2.33	2.08	2.32
police and fire	2.93	2.87	2.92
housing facilities	2.59	3.02	2.82
education facilities	2.86	3.06	3.01
climate	2.32	2.50	2.39

Source: Tong (1978)

Input data needed by a user for the ESMAN is minimal, but the system must contain a relatively large data base, such as weights of locational components and characteristics of available sites. The data base required for the prototype system developed in this study includes nineteen weight tables and a site characteristic table, which includes thirty-four elements for each of three sites.

The prototype system of the ESMAN was run to identify rankings for three sites for a potential food and kindred plant. Using hypothetical data for site characteristics for each of three sites (Table 8.7), the system recommended alternative sites in rank order.

Summary and Conclusions

The performance of an expert system depends on the knowledge it contains. Advantages of using expert systems over other conventional computer programs include the fact that rules that constitute the knowledge base can include both relative and absolute notions, as well as rules of thumb. In the prototype system of ESMAN described above, for example, consider the following comparison:

Table 8.7. Hypothetical Site Characteristics Data.

	Sites		
Characteristics	A	B	C
1 Population	34,389	58,267	45,689
2 Fireman	54	77	68
3 Policeman	62	79	71
4 Waste pick-up	yes	yes	yes
5 Public library	no	yes	yes
6 Number of banks	4	10	7
7 Assets of banks	467897T	975026T	876578T
8 Total tax levy (/100)	7.57	7.05	6.90
9 Full time chamber of commerce executive	no	yes	yes
10 Total labor force	49,805	86,643	76,356
11 Percent unemployed	8.5	5.6	7.3
12 Manufacturing employment	4,320	7.061	6,833
13 Wage rate-assembler-production	9.3	9.6	8.9
14 Federal highway	1	2	2
15 State highway	1	2	1
16 Distance to interchange	0	0	0
17 Freight carrier-interstate	8	13	10
18 Freight carrier-local	19	31	28
19 Package delivery	no	yes	yes
20 Commercial bus service	yes	yes	yes
21 Distance to piggyback station	30	45	50
22 Distance to airport	8	5	12
23 Number of flights per day	21	58	46
24 Longest runway	6500	8000	8000
25 Charter service	no	yes	yes
26 Distance to Atlanta	400	594	630
27 Distance to Chicago	214	136	89
28 Distance to Memphis	632	394	345
29 Distance to New Orleans	768	776	983
30 Cooling degree days	622	789	888
31 Heating degree days	5899	6435	6877
32 Vacant rate	4.2	5.7	6.7
33 Available lot size	37	42	46

```
IF        wage rate of site 1 is greater
            than the average wage rate of the sites,

THEN      wage rate of site 1 is high.
```

Another kind of rule was made up on absolute terms, such as,

```
IF        required parcel size is bigger than
            the largest parcel size available in the site 1,

THEN      there is not a big enough lot for
            the plant in the site 1.
```

However crude it might be, the prototype expert system ESMAN has shown that many urban and regional planning problems that meet the conditions described above may be diagnosed using expert system tools. This way, not only can the knowledge and experience of expert planners be codified in a system that can advise less experienced planners, but such a system will also become an educational tool that benefits planning students.

References

Tong, H. M., 1978. *Plant Location Decision of Foreign Manufacturing Investors,* University Microfilms International, Ann Arbor, Michigan.

U.S. Congress, House, 1977. "International Investment Uncertainty," 9th Cong. 2nd Sess., Washington, D.C., U.S. Government Printing Office.

U.S. Department of Commerce, 1985. *Foreign Direct Investment in the United States,* Office of Business Economics, Washington, D.C., U.S. Government Office.

CHAPTER 9

ESSAS: Expert System for Site Analysis and Selection[1]

Sang-Yun Han and Tschangho John Kim

Urban planning is a multi-dimensional and multi-disciplinary activity embracing social, economic, political, anthropological, and technical factors. As the body of knowledge and the scope of urban activities grow, computers have been increasingly utilized in this field, mostly for the numerical analysis of urban problems.

Solutions for urban and regional planning problems, however, frequently require not only numerical analysis but also heuristic analysis, which in most cases depends on the planners' intuitive judgment. Conventional programming techniques, which mostly deal with numerical analyses of data, lack the capability of incorporating heuristic or qualitative knowledge of planners into problem solving. Expert systems which permit the use of heuristic knowledge or rules of thumb of human experts through computer programs may provide more effective ways of supporting site evaluation tasks of planners.

To examine the usefulness of an expert systems approach to the site evaluation problem, the objectives of this paper are two-fold: (1) to explain the functions and structure of ESSAS: Expert System for Site Analysis and Selection developed for use by the master planners of the Army, and (2) to discuss the implications of applying expert systems to site evaluation problems.

Background of Building ESSAS

The master planners of the Army are responsible for planning future building projects and renovations needed by the people who live and work at the installation, i.e., a military city for people who serve in the Armed Forces, their

---

1   This chapter reprinted with permission from *Computers, Environment and Urban Systems,* 12(4):239-252, copyright 1988, Pergamon Press plc.

families, and civilian employees. Many different types of buildings and support facilities must be planned and built for use by the people living in the installation. Without being supported by effective computer systems, the task of a master planner, who must use heuristic knowledge and numerical routines in evaluating a potential site, is difficult since installations are often the size of large cities. Furthermore, installations contain many different facilities and infrastructures with a limited amount of land and other resources available for construction (see Kim, Han, and Stumpf (1987) for detailed information).

Each type of facility requires certain site conditions to ensure a long economic life and minimal adverse impacts. Environmental concerns as well as water, energy supplies, sewage and other requirements generated by additional people and equipment are important considerations. Remote sites may require a large capital investment to connect buildings to existing utilities and road networks. Zoning considerations and relationships to existing buildings are important as well. The Army has many noisy or hazardous activities, such as weapons training ranges, which must be carefully separated from other uses. Each installation has well defined "Safety Zones" which restrict land uses to assure compatibility.

Master planners are responsible for most site selection tasks which occur in the military setting. They are familiar with the general land use plan for the installation, but may be unable to determine the suitable site without first collecting a substantial amount of data and analyzing them. Types of data include environmental data, cost of infra-structure, surrounding land use, accessibilities, soil conditions and other construction related data.

Previous attempts to develop operations research type models to support the site selection and evaluation tasks have been unsuccessful mainly due to (1) optimal landuse plans devised by these models do not necessarily satisfy critical Army guidelines and regulations which govern the planning, design, construction and siting of each type of building, and (2) most of these models do not incorporate planner's intuitive judgments into the problem solving process.

Expert systems may be regarded as a more relevant approach to the site evaluation tasks of the Army because the major concern of the master planners is not to select a site which minimizes adverse environmental impacts or construction costs or both. Rather, their concern is to select a site which avoids environmental, construction, and other critical problems as well as abides by numerous Army regulations and rules associated with site development.

Thus, the critical part of developing ESSAS was review of the Army regulations, field manuals, and design guides as well as in-depth interviews with the master planners. Specifically, development required encoding site selection rules and the expertise of master planners into a knowledge base. To make the project more manageable ESSAS is currently configured to evaluate the suitability of sites for administrative buildings only. Extensions of this initial knowledge base are now under consideration.

Tasks in Developing ESSAS

The tasks involved in developing ESSAS are discussed in this section. The specific tasks are presented as five steps: (1) identification of important aspects of problem, (2) conceptualization of problem, (3) formalization of knowledge, (4) implementation of knowledge using production rules, and (5) testing of the system.

1. Identification of Problem
    In this phase, the type and scope of the problem, the required resources, and the goals and objectives are identified and then compared to the available resources and the interests of the researchers. The initial problem considered is to model the decision process for site selection for Army training ranges. Since this problem is too broad and requires access to confidential reports on the specification of Army equipment, the scope and complexity of the problem must be narrowed down into a more manageable size. The final subject selected is to model the decision process of the master planner in a specific military installation, and to capture the various expertise needed in selecting a site for administrative building construction. The required data are identified and a point of contact is made with a master planner in Fort Bragg, North Carolina.

2. Conceptualization of Problem
    Concepts, relations, and control mechanisms necessary to describe problem solving in site analysis and selection are decided during conceptualization of problem. Sub-tasks, strategies, and constraints related to this problem solving task are also examined. In order to make data gathering of site characteristics more feasible, the scope of the problem is narrowed down further between four sites of new construction. Determination of the actual site selection process is accomplished at this stage after consulting several experts and many references.

    Another major task involved in this stage is to incorporate Army regulations and guidelines applicable to installation planning for site evaluation and selection. These references include important topics such as environmental issues, construction and safety issues, compatible land uses, building, parking and open space requirements, and utility requirements. The site analysis and selection factors are cross-checked with available expertise in public domain literature.

    The expert system building tool, Personal Consultant plus, a product of Texas Instruments, Inc., is chosen at this stage. It utilizes rule-based knowledge representation methods and backward chaining inference mechanisms and runs on IBM compatible microcomputers.

3. Formalization of Knowledge
    Formalization involves expressing the key concepts of the problem and relations in some structured way, within a framework of the chosen expert

system building tool. It is determined that the expertise in site analysis and selection can be expressed in the form of IF-THEN rules, and the large knowledge base can be effectively organized into sub-areas using FRAMES of Personal Consultant Plus. Frames serve a general organizational purpose by defining problem areas in the knowledge base. The criteria for site analysis are structured into four sub-frames: Environmental, Construction, Safety, and Local Land-Use. Categorization of the site evaluation factors specified in Army regulations and manuals into these four sub-frames requires much effort.

4. Implementation of Knowledge
   During implementation, the formalized knowledge is encoded into a computer program using IF-THEN rules, parameters, and frames. Parameters are specific facts or pieces of information, and rules are logical statements written in IF-THEN format. Rules describe the logical relationship between parameter values. The data structure, inference rules, and control strategies are implemented using Personal Consultant Plus interfaced with Lotus 1-2-3 for data base files and AutoCad for graphic representation of data.

5. Testing
   Testing involves evaluating the performance and utility of the prototype system and revising it as necessary. The system is tested to check if it makes conclusions that experts generally agree to, considers items in the order that the expert prefers, and if the system's explanations are adequate for describing how conclusions are reached.

Knowledge Representation and Inference Mechanism of ESSAS

Like the majority of expert systems developed today, the knowledge representation method used in ESSAS is rule-based. A rule, in a generic form, looks like this:

```
IF <Premise 1> and/or <Premise 2> ...
THEN <Inference 1> and/or <Inference 2> ...
```

Examples of rules used in the ESSAS system are as follows:

```
                          IF
<distance from potential site to species habitat is less than 1000 FT>
                          OR
     <the area of wetlands is more than 20% of total land area>
                         THEN
     <the environmental suitability level of the site is very poor>
```

```
                               IF
<the surrounding land use of the site is more than 50% industrial>
                               OR
<the distance of the site to AIRPORT SAFETY ZONE is less than 1000 FT>
                              THEN
    <the local land-use suitability level of the site is prohibitive>
```

In addition to the rules, the program has access to factual data. The system looks at the factual data such as distance of the site to major utilities, expected capacity of the building, and characteristics of the potential site. Personal Consultant Plus uses parameters to store information or facts in the knowledge base. In this system, there are three ways to obtain factual data by assigning values to a parameter:

1. Provided by the end user: The system can ask a user for the parameter value. For example, a user is asked to provide his preferred density of building and geographic location of a potential site.

2. Obtained from external data file: When the system needs factual data, it reads the external data file and sets the value for a parameter. In this system, site characteristic data, such as land area, distance to major utilities, and presence of incompatible land uses in the surrounding area, are obtained from the external data base.

3. Inferred by the system: The system can generate new data by inferring from the data provided by the user or obtained from the external data base and by using the rules. For example, once the density of building, expected number of occupants, and number of floors in the building are provided by the user, it will calculate the required land size for building construction using an expert rule encoded in the knowledge base, and then determine whether the land area of potential site exceeds the required land size.

In this expert system, the criteria used to evaluate potential sites are organized into four major areas. Thus, the rules and parameters are divided into four groups using sub-frames of Personal Consultant Plus. The rules and parameters which are relevant to all areas are stored in the main frame, while the rules and parameters relevant only to a specific area (e.g. environmental criteria) are stored in a sub-frame. This utility of organizing the knowledge base into sub-frames is quite useful because it enables easy addition of sub-frames later as the subject area covered by the expert system increases.

Once the knowledge base is built through the use of rules, the system tries to obtain conclusions through its inference engine. The inference engine provided by Personal Consultant uses a backward chaining (or goal-driven) inference method, which focuses on finding a rule to provide a necessary parameter value for confirming a hypotheses.

As an example of inference strategy, consider the following goal and rules:

GOAL: To Find RECOMMENDED-ACTION

Rule-1:IF ENVIRONMENTAL-SUITABILITY of the site is ACCEPTABLE, THEN RECOMMENDED-ACTION is USE-THE-SITE.

Rule-2:IF DISTANCE-TO-HABITAT is less than 3000 FT, THEN ENVIRONMENTAL-SUITABILITY is ACCEPTABLE.

Rule-3:IF SLOPE of the site is less than 8 PERCENT and DISTANCE-TO-POWER-SOURCE is less than 1000 FT, THEN CONSTRUCTION-SUITABILITY is ACCEPTABLE.

In seeking a value for RECOMMENDED-ACTION, the system scans the knowledge base sequentially for the existence of the parameter, RECOMMENDED-ACTION in the THEN part of a rule. Rule-1 above is such a rule. Once it locates Rule-1, the system attempts to prove the premises of this rule. The first premise is "ENVIRONMENTAL-SUITABILITY is ACCEPT-ABLE." The system checks working memory for a value for "ENVIRONMENTAL-SUITABILITY" and finding none, scans knowledge base seeking a value for that parameter.

Rule-2 in the example contains a value for "ENVIRONMENTAL-SUITABILITY" in its THEN part. Thus, the system again tries to prove the IF part of this rule. If Rule-2 succeeds, then it sets value for "ENVIRONMENTAL-SUITABILITY," otherwise it continues checking other rules which contain the parameter, ENVIRONMENTAL-SUITABILITY, in their THEN parts. The second condition in Rule-1 is checked in the same manner.

The system can solve multiple goals, one at a time. The list of goals used in ESSAS includes:

1. AREA-BUILDING: Find the land area required for building construction.

2. AREA-PARKING: Find the land area required for parking lots.

3. AREA-OPEN-SPACE: Find the land area required for open space provision.

4. SURROUNDING-LAND-USE: Provide surrounding land use information for the site selected by the user in a digitized map form if the site is recommended.

5. CRITICAL-PROBLEM: Find the critical problems expected in using the site.

6. RECOMMENDED-ACTION: Recommend a desired action to the user regarding the use of the site.

Function of ESSAS: How it Works

As summarized in Figure 9.1, ESSAS initiates consultation by the user's inputs and then tries to reach conclusions at the end of the user consultation. To reach conclusions, it has several goals to meet, such as: 1) to provide the user with the information inferred by the system regarding the use of the potential site, and its suitability in terms of several different criteria, 2) to predict some possible serious problems for potential site use, 3) to provide recommended system action regarding the use of the potential site.

The following is an explanation of the consultation procedures used in ESSAS:

1. The USER is asked to provide the expected capacity of building in terms of number of occupants and vehicles, and the preferred density of buildings.

2. The SYSTEM calculates the spatial requirements of the site for building, parking, and open space construction, based on the user inputs and ESSAS knowledge on estimating spatial requirements.

3. The USER is asked to select a site for evaluation from a Fort Bragg land-use map provided by the system. This allows ESSAS to reach the external data base for the user selected site and know the characteristics of that site. All potential sites are pre-specified in the map and the external data base maintains site characteristic data for these potential sites.

4. Once the system has all the information it can get from the user and external data base, the SYSTEM starts reasoning to reach conclusions about the goals specified utilizing its knowledge base. First, it evaluates the site using critical rules in the knowledge base. An example of a critical rule is "IF the land area of the site is less than the required land area, THEN reject the site." If a site under evaluation doesn't satisfy any of the critical rules, the system immediately reaches a conclusion about the use of the site (i.e. DON'T USE THE SITE) and lists the critical problems detected by the system.

5. When a potential site passes all the critical rules, the SYSTEM enters the first sub-frame, ENVIRONMENTAL, and evaluates the site using environmental criteria, such as, impacts on wetlands and endangered species habitat. The SYSTEM reports serious environmental problems as well as positive features regarding the use of the site. It also tries to determine an ENVIRONMENTAL SUITABILITY LEVEL of the site with which the system determines overall suitability of the site and the action recommended by the system at the end of consultation. Seven different levels of suitability are used here: EXCELLENT, VERY GOOD, GOOD, ACCEPTABLE, BAD, VERY BAD, and PROHIBITIVE. The system determines the suitability level based on the degree of

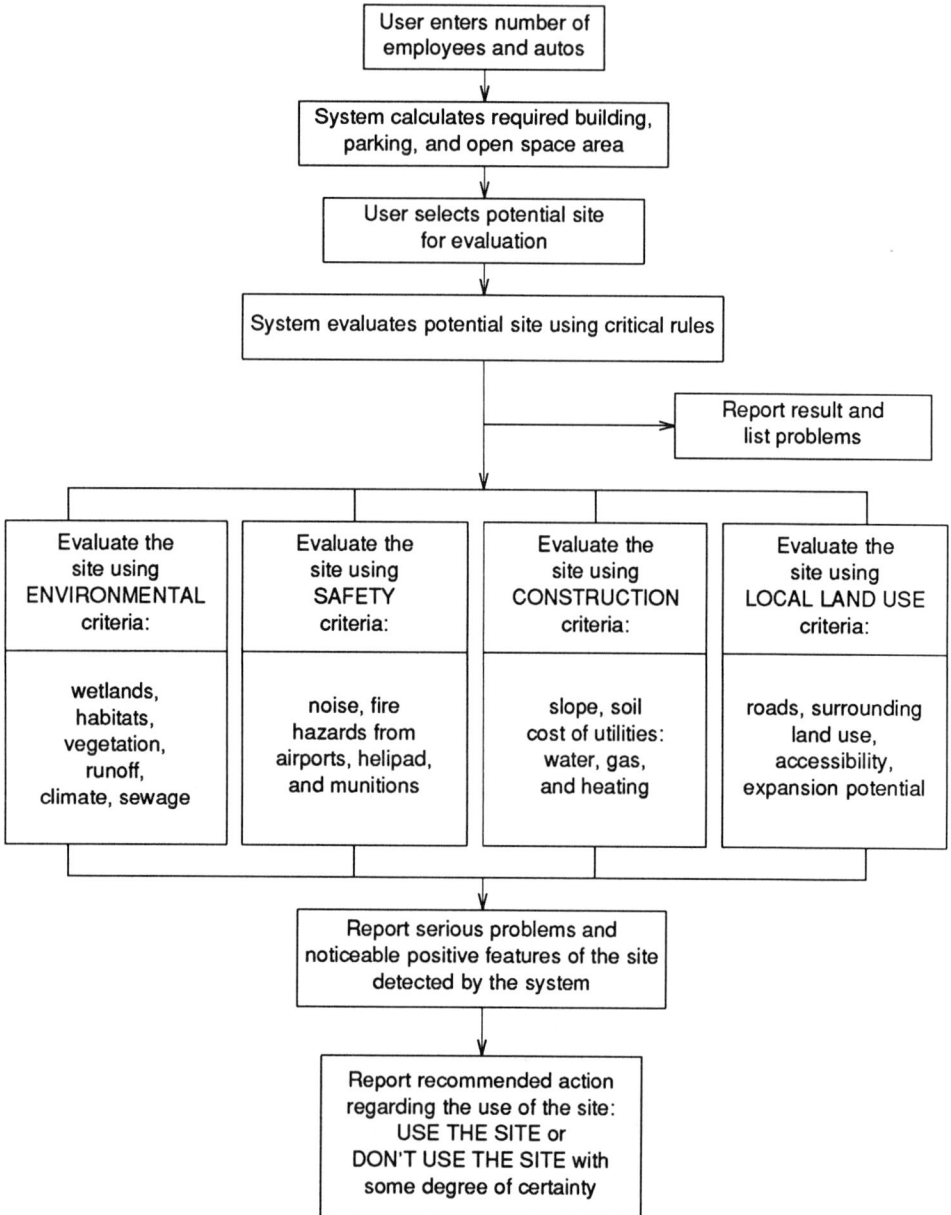

Figure 9.1. Schematic structure and function of ESSAS.

environmental problems it detects.

6. After it completes solving the goals of the ENVIRONMENTAL frame (i.e. to find environmental problems, and environmental suitability level), the SYSTEM enters the SAFETY sub-frame and evaluates the site using safety and health criteria such as noise level, expected hazards from air-port, helipad, and munitions storage. Using its knowledge base, it reports safety-related problems as well as attractive features of the site and deter-mines a SAFETY SUITABILITY LEVEL.

7. The SAFETY sub-frame continues this process for the rest of sub-frames (CONSTRUCTION and LOCAL LAND-USE), and then returns to the main frame.

8. In the main frame, the SYSTEM reports the area needed for building, parking, and open space, and retrieves from its data base the land-use map of the site which shows the area within 1/2 mile radius of the site for user information. Finally the SYSTEM, based on the suitability levels from the four sub-frames, determines the action recommended by the system regarding the use of its conclusion: USE THE SITE or DON'T USE THE SITE with a determined certainty factor, and then summarizes the suita-bility of the site in terms of environmental, safety, construction, and local land use criteria. In determining the final action, the system relies on a set of rules which decide the final action based on the suitability levels from four sub-frames. For examples of such a rule:

```
                        IF
    <ENVIRONMENTAL-SUITABILITY is EXCELLENT>
                       AND
        <no PROHIBITIVE in other criteria>
                      THEN
                 <USE THE SITE>

                        IF
        <any criterion is PROHIBITIVE>
                      THEN
              <DON'T USE THE SITE>
```

9. The USER can continue evaluating other sites by selecting any site from the land use map with different user inputs. The USER can ask how the system reaches certain conclusions by selecting HOW from the menu. The SYSTEM will list the rules it used to generate rules. For example, when the user asks how the total land area required for building construc-tion was determined to be 1,000,000 square feet, the system will list the rules containing formulae calculating spatial requirements and its refer-ence sources.

Future Extensions of ESSAS

The expert system ESSAS is a medium sized research prototype system with 240 rules, designed to solve a portion of the problem undertaken to suggest that the approach is viable and full system development is achievable. To have practical value, an extensive field test is required to produce a more reliable system that better addresses the needs of the end-user.

One possible extension to ESSAS includes embedding some mechanism to resolve conflicting knowledge in the knowledge base. Since the expertise encoded into expert systems is normally obtained from multiple knowledge sources, it is fairly typical that several conflicting or different rules apply to the same problem in the reasoning process. Because conflict resolution strategies are not explicitly specified, currently ESSAS applies the first rule encountered in the reasoning process when conflicts arise.

More advanced ways of resolving conflicts may be employing meta-rules which are explained as high-level rules about how to use rules or integrating analytical models into expert systems to resolve conflicts in the knowledge obtained from rules-of-thumb. The first approach is explained by Waterman et al. (1983), D'Angelo et al. (1985), and Han (1988) and the second approach is proposed by Han (1988).

Another possible extension includes integrating database management systems into ESSAS. It has been noted that storing large sets of data in rule format consumes a large amount of computer memory and slows down processing speed of the system. It is learned from ESSAS that to be a practical tool, expert systems need to be supported by relational database systems which effectively manage and process factual data. To save time and resources while increasing the productivity of the system, ESSAS might be reconfigured to use existing data base systems. Since the use of digitized maps can be useful for many planning problems including site selection problem, it will be also worth attempting to develop expert systems which read information directly from digitized maps and incorporate it into the knowledge base. Although ESSAS is currently designed to use maps digitized using AutoCad during consultation process, the interaction between ESSAS and the computer mapping program is limited. Coupling Landtrak or any other geographic information systems package into ESSAS may create more practical expert systems.

Implications of Applying Expert Systems to Site Evaluation Problems

In urban planning, optimization models and suitability analyses using overlay manipulation techniques have been dominant methods for the site selection and evaluation tasks. Some examples of optimization models used in this area include; Bammi et al. (1976) which is formulated to minimize environmental impacts; Kim (1978) which devises optimal zoning schemes minimizing costs associated with commodity production, transportation, and export; and Barber (1976) which provides a multiple-objective optimization model minimizing land development costs and energy consumption in transportation and maximizing residential accessibility. On the other hand, the suitability analyses use overlay manipulation techniques which combine a set of different map overlays to produce a composite suitability map as illustrated by McHarg (1971) and Steinitz et al. (1976).

What is the importance of expert systems when these tools of optimization modeling and overlay analysis are already available? It may be argued that expert systems provide an alternative way of supporting site selection and evaluation tasks in situations; first, where traditional approaches are inadequate, and second, where traditional approaches are insufficient by themselves. To illustrate the first situation, consider the site evaluation problem dealt with by ESSAS. The users of ESSAS, i.e., the master planners of the Army, do not have clear objectives to be optimized in the site evaluation process. Rather, they want to select a site which complies with the Army regulations, field manuals, and directives from high-ranking officials, and conforms to the planners' qualitative knowledge on site evaluation.

Obviously, optimization models are inadequate in this situation. In addition, interpreting the Army regulations regarding site evaluation, for example, is not a matter of numerical manipulation of data of most algorithmic programming, but a matter of symbolic manipulation of data which is well suited for expert system applications.

In a situation where problem-solving objectives and strategies are not fixed, because, for example, the directives from decision makers frequently change and the knowledge of planners are enhanced through their new training and experiences, the expert systems approach may be more suitable than the algorithmic programming approach. This argument is based on the distinctive nature of expert systems, that is, the separation of domain specific knowledge from the inference engine which is the reasoning mechanism for applying that knowledge to an instance of problem. In traditional algorithmic programming, knowledge about problem domain which is normally expressed in the form of mathematical equations is mixed with strategies which control the flow of that knowledge using procedural statements, such as DO-WHILE, GOTO, STOP, IF, and ENDIF. While efficient search strategies including depth-first, hill-climbing, and other heuristic search strategies are common interests of both expert systems and algorithmic programming, the development process of any expert systems mostly focuses on the development of the knowledge base, leaving the search tasks to the built-in inference engine provided by the expert system development shells.

There is an important implication of separating knowledge base from inference mechanism in ESSAS. Most of all, it is easy to upgrade or modify the system by means of adding, deleting, and changing production rules in the knowledge base without worrying about the search mechanism. As previously illustrated, the production rules are the simple IF-THEN rules used to represent the knowledge in ESSAS. The following rule,

```
IF proposed use of site is administrative buildings
THEN the site must be outside the safety zones
```

can be easily changed to,

```
IF proposed use of site is administrative buildings
THEN site must be outside the safety zones and
        within 1000 feet of the commanding offices.
```

The separation of knowledge from inference engine enables the easy modification of expert systems, which in turn enables repetitive applications of ESSAS with minor upgrades to essentially similar problems.

The experiences from ESSAS indicate that expert systems may also be used to provide leverage to the suitability analysis which employs traditional overlay manipulation techniques. As pointed out by Hopkins (1977 and 1979), the "rules of combination" method which is based on the knowledge of experts can be a valid and effective way of combining overlays, because it "states in words the combination of levels of impacts that are determined to be preferred to other combinations" (Hopkins, 1979).

Because the process of any site evaluation tasks inevitably deals with several factors and criteria (safety, environmental, construction, and local land use criteria and numerous factors in ESSAS), it is essential for any computer based systems developed for site evaluation purpose to incorporate valid methods for the combination of different factors. The critical deficiencies of some commonly used overlay combination methods include adding ordinal scale numbers or assuming independence between different factors when they are interdependent (Hopkins, 1977).

To some degree, ESSAS implements "rules of combination" in its knowledge base. For example, consider the following two rules:

```
Rule 1:
IF erosion potential is high and wetland is not present
THEN environmental suitability is poor

Rule 2:
IF erosion potential is low and wetland is present
THEN environmental suitability is very poor.
```

These two rules imply that the combination of "high erosion potential" and "no wetland" is preferred to the combination of "low erosion potential" and "existence of wetland". This rule of combination is based on the knowledge of experts that preventing erosion problems is less costly that saving wetlands. Although the use of heuristic rules of combination is not fully utilized in ESSAS, certainly ESSAS sheds light on how different decision factors are effectively combined using the knowledge of experts and is efficiently encoded into the knowledge base of expert systems.

In summary, there are at least two important implications of applying expert systems in urban planning. First, the use of expert systems broadens the spectrum of planning problems which can be supported by computer based systems. Expert systems can be useful in a situation where traditional problem-solving approaches are inadequate either because the problem to be addressed is not an optimization problem or because the problem-solving strategies frequently change. Secondly, expert systems may supplement existing techniques used for site evaluation tasks when the existing techniques are not sufficient by themselves. The use of "rules of combination" in ESSAS is proposed as an example which supplements traditional overlay techniques.

Concluding Remarks

The process of building an expert system is iterative and involves much trial and error. The prototype expert system ESSAS attempts to synthesize multi-disciplinary activities of urban and regional planning into a comprehensive decision-making tool for site analysis and selection, as an alternative to the conventional techniques. It demonstrates that expert systems can be useful tools to professional planners for their site evaluation tasks by performing difficult or tedious tasks including reviewing critical regulations and rules governing environmental, safety and health, construction and local land use factors.

The purpose of this paper is to provide a basis for speculating on the ways in which expert systems can be effectively used for professional planners. It is hoped that the development of ESSAS will shed light on the issues involved in building a detailed and comprehensive expert system to aid professional planners.

References

Bammi, D., D. Bammi and R. Paton, 1976. "Urban planning to minimize environmental impact," *Environment and Planning A*, 8:245-259.

Barber, G.M., 1976. "Land-use plan design via interactive multiple-objective programming," *Environment and Planning A*, 8:625-636.

D'Angelo, Antonoi, Giovanni Guida, Maurizio Pighin and Carlo Tasso, 1985. "A mechanism for representing and using meta-knowledge in rule-based systems," in *Approximate Reasoning in Expert Systems*, ed. Gupta et al., New York: North-Holland.

Goodall, Alex, 1985. *The Guide to Expert Systems*, New York: Learned Information, Inc.

Han, Sang-Yun, 1988. "Design, Implementation and evaluation of an integrated decision support system: Strategies for resolving conflicts in land use planning," Proposal for Ph.D Dissertation. Department of Urban and Regional Planning, University of Illinois at Urbana-Champaign.

Hart, Anna, 1986. *Knowledge Acquisition for Expert Systems,* New York: McGraw-Hill Book Company.

Hopkins, Lewis D., 1977. "Methods for generating land suitability maps: A comparative evaluation," *Journal of American Planning Association,* October:386-399.

Hopkins, Lewis D., 1979-1980. "Land suitability analysis: Methods and interpretation," *Landscape Research,* 5:8- 9.

Kim, T. John, 1978. "A model of zoning for a metropolis." *Environment and Planning A,* 10:1035-1045.

Kim, T. John, Sang-Yun Han, and Annette Stumpt, 1987. *Developing expert system for site anylsis,* Report to the U.S. Army Construction Engineering Research Lab (DACA 88-86-D-0006-40)

McHarg, Ian L., 1971. *Design with Nature,* New York: The Natural History Press.

Ortolano, Leonard and Catherine D. Perman, 1987. "A Planner's Introduction to Expert Systems," *Journal of the American Planning Association,* 53.

Steinitz, Carl, Paul Parker and Lawrie Jordan, 1976. "Hand-drawn overlays: Their history and prospective uses," *Landscape Architecture,* September:444-455.

Waterman, Donald, F. Hayes Roth and D.B. Lenat, 1983. *Building Expert Systems,* Massachusetts: Addison-Wesley Publishing Company.

Waterman, Donald A., 1986. *A Guide to Expert Systems,* Massachusetts: Addison-Wesley Publishing Company.

Wiltshaw, D.G., 1987. "Expert Systems and Land Development Expertise," *Land Development Studies,* 4.

CHAPTER 10

Landfill Site Selection[1]

Shahrokh Rouhani and Roozbeh Kangari

Landfills are typical waste disposal facilities in urban areas. The Sanitary Landfill Manual of Practice, prepared by the American Society of Civil Engineers (ASCE, 1959) defines it: "as a method of disposing of refuse on land without creating nuisances or hazards to public health or safety, by utilizing the principles of engineering to confine the refuse to the smallest areas, to reduce it to the smallest practical volume, and to cover it with layers of earth at the conclusion of each day's operation or at such more frequent intervals as may be necessary." This operation requires systematically depositing, compacting, and covering the wastes. For example, usually 6 to 12 inches of earth are placed over each 2 feet of compacted fill. The top earth cover should have a minimum designated depth and be grassed to prevent erosion. The projected land use of a landfill may be a park with recreational facilities that are not affected by gradual subsidence of the ground surface. In general, landfills offer an economic alternative to the problem of the disposal of wastes.

Prior to designating an area as a landfill, detailed comprehensive studies must be conducted to select the most suitable location for such a facility. The technical process of site selection considers potential migration routes of the wastes, waste characteristics, potential targets at risks, planned features of the landfill, along with a comparative cost-benefit analysis.

Potential migration routes involve the study of contaminant movement away from the disposal site by ground waters, surface waters, and air routes. The movement of wastes depends on physiographic and climatic characteristics of the site. It also depends on the waste characteristics, which can be quantified by its toxicity, persistence, ignitability, and reactivity. Potential targets include population centers at the vicinity of the landfill, nearby sensitive ecological systems, and critical habitats.

---

1   This chapter reprinted by permission of the publisher from "Landfill Site Selection: A Microcomputer Expert System," by S. Rouhani and R. Kangari, *Microcomputers in Civil Engineering*, 2:47-53. Copyright 1987 by Elsevier Science Publishing Co., Inc.

The planned features of the facility are also the important factors in such an investigation. These are represented by means to minimize or prevent a contaminant from entering into a migration route and reaching a target. Finally, site selection should include a cost-benefit analysis to determine the economic efficiency of the proposed facility. For more information readers are referred to (Nobel, 1976).

The above site selection process is basically composed of four parts: (1) available soil/water quality, physiographic, climatic, and socioeconomic data; (2) a set of explicit algorithmic models and analyses based on rules which are implicitly embedded in these models; (3) a set of rules, standards, regulations and guidelines set by the federal and state agencies; and (4) explicit expert opinions and judgements. Synthesis and analysis of these components require a significant amount of expert inputs, which in addition to explicit heuristic rules and opinions, include such tasks as: designing the general approach and a uniform ranking procedure, identifying constraining regulations, selecting the appropriate algorithmic models, preparing the required model inputs' analyzing model outputs, and finally, selecting or recommending a specific site and size for a landfill facility. The expert plays a central and vital role in this process, as shown in Figure 10.1.

To accomplish the above tasks an expert has to rely on his or her judgements and expertise. This expertise is the product of an iterative and explorative learning process, in which the expert evolves his or her judgement into a set of explicit rules. These rules can be denoted as empirical knowledge which includes expert opinions and inferences, and rules of thumb (Kangari and Rouhani, 1986).

An Expert System Framework

It is very difficult if not impossible for various teams composed of engineers and scientists to possess the broad expertise and knowledge required to perform such site selections in a uniform and standard manner. For instance, model developers are only concerned with the explicit algorithmic parts of their work. So there is a tendency to ignore the significance of empirical knowledge in the design of the entire program. As a result, model users are frequently faced with problems, such as, how to interpret and prepare data, which parameter values to choose, how to fit a slightly different environment to the model, what to include and what to leave out and how to analyze and interpret the output. In many instances, these problems have made it impossible for anyone but the model developer to explore his or her empirical knowledge, and use the model successfully.

One of the reasons for this deficiency is the fact that expert knowledge is not necessarily algorithms or effective procedures of computer sciences. Usually, empirical rules are not complete, or unique, or in a specific sequence to be programmed in traditional algorithmic frameworks. A solution to this problem

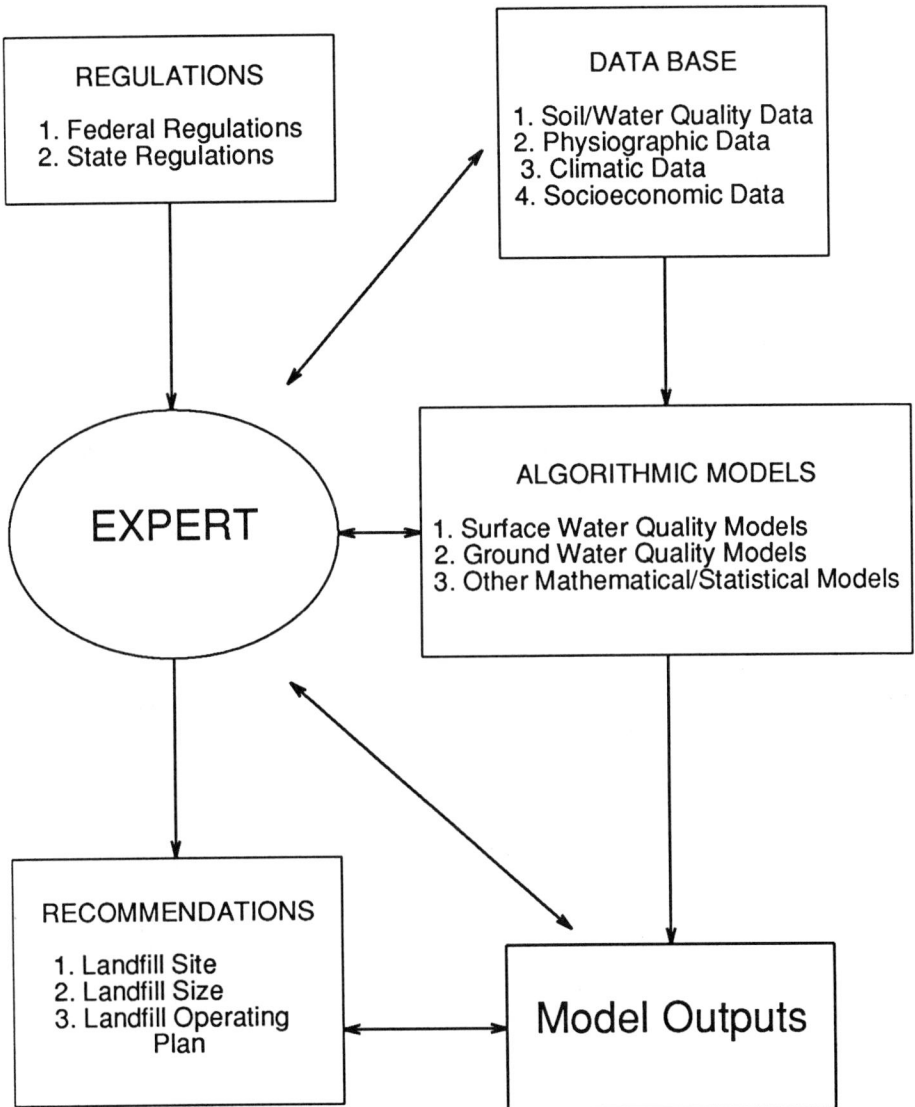

Figure 10.1. A typical landfill design process.

is created by recent advances in Knowledge-Based Expert Systems (KBES). Using elements of heuristic programming, we can incorporate our empirical and expert rules into our decision making process to yield a more efficient planning procedure (Waterman, 1986).

A knowledge-based expert system is a computer program which automates the delivery of advice and expertise in a specific problem area. The knowledge is typically represented in three ways, which are, production rules, frames, and semantic nets. In our prototype model the production rules of IF/THEN/ELSE form seem to be fully adequate for the presentation of knowledge. These rules are a formal way of specifying how an expert reviews a condition, considers various possibilities, and recommends an action, which are expressed as:

```
IF (premise), THEN (action); or,
IF (condition), THEN (action).
```

Knowledge plays an important role in an expert system. In many cases it is difficult to obtain knowledge. Our experience shows that for building a knowledge base we must work with the source of knowledge (e.g., an expert) in the context of solving particular problems, rather than directly posing questions about rules associated with them. In this particular study we have utilized a U.S. EPA document, 1984 as the source of our knowledge, as described in the next section.

Based on such a framework, we have divided the landfill site selection into two basic parts: the support elements and the core system, as depicted in Figure 10.2. The objective of the first part is to collect information from various sources, such as: a domain expert, specification and guidelines, investigation reports, and other appropriate sources. The second part consists of the core system, which can combine the empirical knowledge with the data base and algorithmic models to provide the user with the necessary advice and explanations. In our prototype model we have limited our core system to the knowledge base, the inference engine, the explanation system, and the user interface.

Our landfill site selection model is programmed with the aid of Insight 2+. This is a microcomputer expert system shell program, written in PASCAL, based on backward chaining. Knowledge is represented in the IF/THEN/ELSE production rules with confidence factors that can be assigned either automatically, or in response to the user's request. Insight 2+ can activate outside algorithmic programs, and has mathematical computation capabilities (Level Five Research, Inc., 1985).

Prototype Model Description

In this study we have considered a site selection procedure for a landfill facility with potential ground water migration routes. This system is intended to be used as a consultant to assist waste disposal managers in the preliminary phases of site selection. Increase in efficiency, improvement and uniformalization of selection strategies, and automated record keeping are among the expected

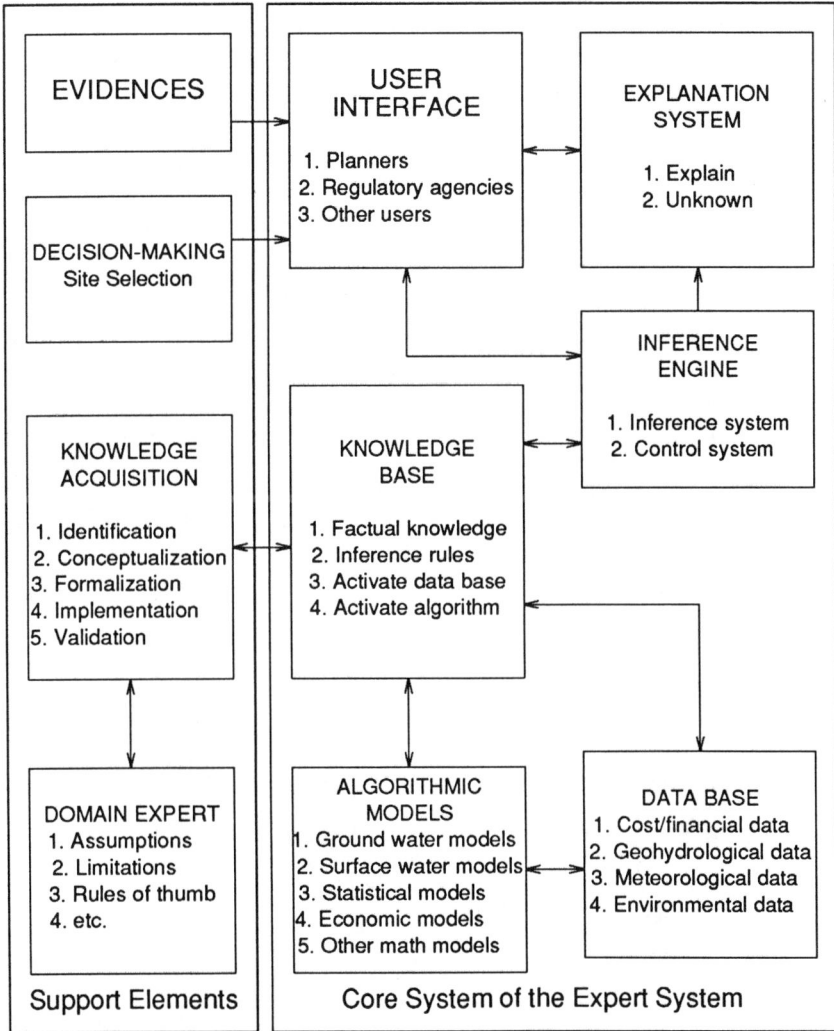

Figure 10.2. Basic structure of the expert site selection system.

benefits of such an expert advisory system.

The above model is based on production rules derived from U.S. Environmental Protection Agency (1984). This U.S. Environmental Protection Agency document provides a ranking procedure for remedial actions for uncontrolled hazardous waste sites, as required by the Comprehensive Environmental Response, Compensation, and Liability Act of 1980 (CERCLA). The ranking criteria are based on: "relative risk or danger, taking into account the population at risk; the hazardous potential of the substances at a facility; the potential for contamination of drinking water supplies, for direct human contact, and for destruction of sensitive ecosystems; and other appropriate factors." These rules and criteria are developed by the MITRE Corporation, and extensively reviewed by U.S. Environmental Protection Agency personnel, state officials, and interested parties in the private sector.

For our case study we have transformed the above expert rules into the IF/THEN formats without any major difficulty. These rules are utilized as the basis to rank the potential landfill locations to identify the more suitable sites for further intensive studies. Four main factors are identified as: (1) ground water route characteristics, (2) waste characteristics, (3) facility characteristics, and (4) targets at risk. The model assigns a score to each factor that will be eventually used to determine an overall score of the site for ranking purposes. The decision making process is shown in Figure 10.3. These factors are:

1. *Ground water route characteristics.* The rules concerning the ground water route involves variables such as, the depth of the lowest point of the landfill to the highest seasonal level of the saturated zone of the aquifer of concern, net seasonal rainfall, permeability of the vadose zone, and the physical state of the waste at the time of disposal. These aspects are formulated into four groups of rules, which include production rules similar to the following two examples:

   RULE 103:

       IF the depth of aquifer of concern with respect
          to the lowest point of the proposed landfill
          is between 21 to 75 ft;

   THEN the condition is relatively risky
        with an assigned value of 4.
   and,

   RULE 132:

       IF the waste is solid,
      AND it is unconsolidated;
       OR the waste is solid,
      AND it is unstabilized;

   THEN the condition is relatively safe
        with an assigned value of 1.

   At this stage, based on the answers provided by the user, appropriate rules will be executed. The sum of executed assigned values will be denoted as the ground water route characteristics score (S_{gr}).

2. *Waste characteristics.* To determine the waste characteristic score (S_{wc}), the model considers the toxicity and persistence of each potential waste

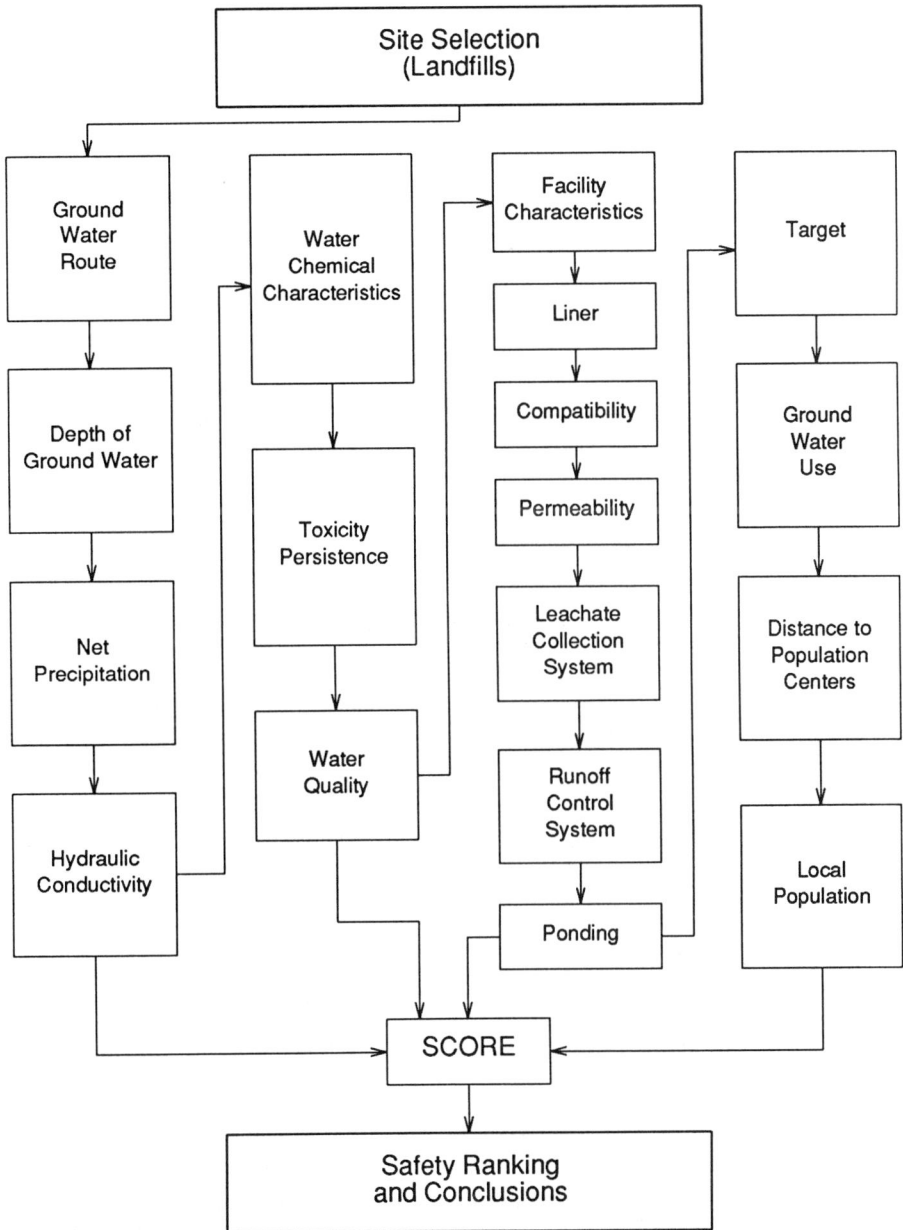

Figure 10.3. Inference net of the expert site selection model.

substance, along with its projected disposed quantity. The substance with the highest score is then selected as the representative waste material. These rules include:

```
RULE 206:

  IF the substance is relatively toxic,
 AND it is highly persistence;
  OR the substance is toxic,
 AND it is persistence;

THEN the condition is risky
        with an assigned value of 15.
```

or,

```
RULE 216:

  IF the waste quantity is between 251
     and 625 tons per cubic yard;

THEN the condition is moderately risky
     with an assigned value of 5.
```

The highest sum of the two executed assigned values among all projected waste substances is then denoted as the waste characteristics score.

3. *Facility characteristics.* This stage includes the expert rules about the planned containment characteristics of the proposed landfill facility. The following is one of these rules:

```
RULE 302:

  IF the landfill has a non-permeable liner,
 AND it has a compatible liner,
 AND it has no leachate collection system,
 AND it has no run-on control system,
 AND it precludes ponding;

THEN the condition is relatively safe
     with an assigned value of 1.
```

The executed assigned value is defined as the facility characteristics score (S_{fc}).

4. *Targets.* The corresponding rules at this stage characterize the level of risk to potential targets. They include rules about the potential use of aquifer of concern, the vicinity of nearest production wells, and nearby population served by the ground water. They include three groups of rules such as:

```
RULE 403:

   IF the ground water is used for drinking,
  AND there is an alternate source;
   OR the ground water is used economically,
  AND there is no alternate source;

 THEN the condition is relatively risky
      with an assigned value of 2.
```
or,
```
  RULE 416:

     IF the distance to the nearest production
        well is between 1 to 2 miles,
    AND the population is between 101 to 1000;
     OR the distance to the nearest production
        well is between 2 to 3 miles,
    AND the population is between 1001 to 3000;

   THEN the condition is relatively safe
        with an assigned value of 12.
```

The sum of the three executed assigned values will then yield the target score (S_t).

The above model contains 76 rules, and considers 99 different facts, which are elements that describe the domain of our knowledge concerning preliminary landfill site selection. The expert system shell program converts all these rules into relevant questions. The user has to select appropriate answers to sixteen such questions, covering the following aspects: (1) the depth of the aquifer, (2) local net precipitation, (3) the hydraulic conductivity of the vadose zone (4) the physical state of waste at the time of disposal, (5) the toxicity of waste, (6) the persistence of waste, (7) projected waste quantity, (8) permeability of the proposed liner, (9) compatibility of the proposed liner with waste, (10) proposed landfill leachate control system, (11) surface ponding characteristic of the proposed landfill, (12) proposed landfill run on control system, (13) present type of ground water use, (14) availability of alternative water resources, (15) distance to the nearest production well, and (16) the size of the local population served by the ground water. After this stage the expert system matches the selected answers with each rule, executes the appropriate ones, and calculates the four partial scores. It then derives the overall site score, as follows:

$$S_s = (S_{gr} \times S_{wc} \times S_{fc} \times S_t) \times (100/57330)$$

The overall site score is a value between 0 and 100. It can be utilized as a measure for the relative ranking of a site. In a preliminary study, sites with high scores should be eliminated as potential landfills, while locations with low scores can be targeted for more intensive studies. These scores can also be used as uniform measures for the safety of different sites.

There are also HELP options available to the user to better explain a question. These options are designed to define variables involved in any question, so the user can select the appropriate answers. Furthermore, the Insight 2+ report functions are capable of providing additional supplementary information. For instance, a WHY option describes the reason for posing a specific question. This option may be followed by the REPORT option, which provides the list of executed rules in order to define the system's line of reasoning. There are also other options available which supply such information as, the current rule, simple and numeric facts, and intermediate and final conclusions. These report functions help the user to study the general expert approach to the problem of landfill site selection, and gain synthetic experience. The above explanation facilities constitute one of the fundamental differences between the traditional computer programs and expert systems.

Conclusions

The above prototype model illustrates that expert systems can act as a storage for our empirical knowledge. They can also be used as a means for testing the available empirical knowledge. Indeed, in initial runs of this program an ambiguity in stated facility characteristics rules in U.S. Environmental Protection Agency, 1984, was discovered. The ambiguity was due to an incomplete designation of assigned values for certain combinations of landfill features. For instance, there was no assigned value for a proposed landfill with moderately permeable liner and a run on control system. In another case, the stated rules seem to suggest that a landfill with moderately permeable, compatible liner can acquire an assigned value of either 2, or 3. Consequently, for such proposed landfills whose features could not be matched exactly with a stated rule, the expert system would assign an incorrect S_{fc}, which in turn would result in an underestimated or overestimated overall site score. This problem was corrected by modifying and expanding these rules based on the stated framework in U.S. Environmental Protection Agency, 1984.

Unlike an algorithmic computer program, an expert system can easily be updated. In fact, there is no need for rewriting the program to include a new rule. This feature makes the expert system an ideal tool for incremental programming, which is essential in early stages of modeling of a complex process. Furthermore, its rules can be written in natural language which makes the program readable and understandable. Finally, its explanation facilities are capable of offering valuable information to the user, which can lead to a more efficient planning procedure.

However, due to its flexible search approach, an expert system is usually slower and less efficient than an algorithmic program. Thus, for those problems that have a clear structure, traditional programs are more suitable. In many instances, the initial programming of a process based on an expert system has illustrated a well-structured line of reasoning, which in turn, has led the model developers to more efficient algorithmic programs.

The authors would like to state that their prototype expert system is available to interested readers upon request. The requirement for running this program is the Insight 2+ software (Level Five Research, Inc., 1985). The Insight 2+ system is designed to run on the IBM-PC, XT, AT, and their compatibles. It requires two double-sided floppy disk drives or a hard disk drive and a minimum of 448 Kb of random access memory (RAM). The recommended configuration for using Insight 2+ is 512 Kb of RAM and a hard disk drive. The user must have PC-DOS or MS-DOS version of 2.0 or later operating system. This program is also available on the latest version of Insight 2+, known as Level 5 (Information Builders, Inc., 1988).

This work was supported partially by a grant from the National Science Foundation. The authors also acknowledge the contributions of Nicholas Latoussakis, a graduate research assistant at the School of Civil Engineering, Georgia Institute of Technology.

References

American Society of Civil Engineers, 1959. *Sanitary Landfills,* ASCE Manuals of Engineering Practice, No. 39.
Information Builders, Inc., 1988. *Expert Systems Software, Level 5*, User Manual, New York, NY.
Kangari, R., and Rouhani, S., 1986. "Expert systems in reservoir management and planning," *Proceedings of Water Forum '86*, ASCE, Long Beach, CA, August.
Level Five Research, Inc., 1985. *Insight 2*, User Manual, Melbourne Beach, FL.
Nobel, G., 1976. *Sanitary Landfill Design Handbook*, Technomic Publishing Co., Westport, CT.
U.S. Environmental Protection Agency, 1984. *Uncontrolled Hazardous Waste Site Ranking System*, A User Manual, HW-10, Washington, DC.
Waterman, D.A., 1986. *A Guide to Expert Systems*, Addison-Wesley Publishing Co., Reading, MA.

Expert Systems in Environment, Conflict Mediation and Legal Disputes

Introductory remarks by Lyna L. Wiggins

The chapters in Part Three of this book were designed to address different aspects of a single task, that of site selection. In Part Four we have selected a group of systems that focus on the common theme of "decision-makers' aids." The three expert systems included here address problems that are quite diverse in their substantive area tasks. However, in each case they focus on the decision-maker support aspects of the tasks. They are designed to either help a decision-maker frame and think about "what if" questions, to help train inexperienced decision-makers, or to aid a decision-maker/mediator in a real-time meeting situation.

In Chapter 11, Han and T.J. Kim describe a prototype expert system designed to simulate world coal trade patterns based on rules of thumb about the operation of international markets. An expert system is appropriate for this task because of its effectiveness in modeling human behavior in problem solving. The decision-making procedures of coal buyers often involve heuristics that can be explained as "rules of thumb," and an expert system provides a technique for modeling this judgmental and heuristic knowledge. Examples of such heuristics in this task include decisions related to "diversity of supply" concerns and to changing political situations between foreign countries. The Expert System for Global Coal Trade (EXCOTRA) is a computer system that enables decision makers in coal marketing to experiment with different variables in order to answer "what if" questions regarding major future events which may affect the world coal market (e.g., miners' strikes, trade sanctions, etc.). In particular, the system is designed and implemented in the context of the Asian coal market, using Korea as the test country.

In Chapter 12, M.P. Kim and Adams discuss an expert system for claims guidance that has been developed at the U.S. Army Construction Engineering Research Laboratory. The government has goals of minimizing claims by contractors due to unknown subsurface conditions at construction work sites, and of minimizing the risk to contractors of such unanticipated conditions. There are two stated objectives for this expert system. The first objective relates to assisting less experienced personnel in making decisions, and focuses on providing an inexperienced project engineer with prelegal assistance in the analysis of potential claims from construction contracts. The second objective relates to training, and focuses on providing a system that may serve as a training device for new personnel in field offices, familiarizing them with the related legal issues. This chapter describes the development of the first module of the Claims Guidance System (CGS) for analyzing "Differing Site Condition" (DSC) claims. The system was developed using an expert system shell, Personal Consultant Plus by Texas Instruments, and implemented on a microcomputer.

In Chapter 13, Lee and Wiggins describe the design and implementation of a computer-based decision support system (MEDIATOR) that uses various methodologies including decision analysis, multivariate analysis, expert systems, and dispute resolution techniques. The system is designed to aid a mediator in a real-time facilitation of a meeting that is concerned with an environmental dispute. In particular, the case used for the prototype system is a dispute over the weights on site selection factors for an energy facility in California. Four research steps were followed in the creation of this system: (1) an articulate expert was identified; (2) the environmental dispute mediation task was narrowed and formalized; (3) a prototype system was developed; and (4) a system validation exercise was completed using a substantially different type of siting conflict. It is argued by the authors that by combining the computational power of conventional computer-based analytical programs with the problem- solving potential of heuristic-oriented systems, that a broad-based system that draws a more complete picture about the problem may be devised. In AI jargon, these types of hybrid programs are called "coupled systems." The system was developed using an expert system shell, NEXPERT, coupled with other software application packages, and implemented on a Macintosh II personal computer.

CHAPTER 11

EXCOTRA: An Expert System for Global Coal Trade

Sang-Yun Han and Tschangho John Kim

In response to the oil embargoes during the 1970s many of the non-oil producing countries have adopted new energy policies aimed at decreasing their dependency on imported oil supplies. These policies in turn generated a new demand for steam coal. Although the increase in coal demand has been slower than initially expected due to low crude oil prices, it is anticipated that the international coal market will continue to be robust over the following decade (Kim and Suh, 1986).

As with any international commodity, a key question is how to explain observed trade patterns and predict future ones as well. Further, it becomes crucial for government officials and private sector planners in coal producing states to understand how government policies in importing countries can influence that trade. The prototype expert system, EXCOTRA: Expert System for Global Coal Trade, is designed and implemented within this context and with the intentions of modeling coal demand patterns in the Asian market, especially for Korea.

The objective of this chapter is two-fold: 1) to discuss the suitability of the expert systems' techniques for building simulation models, and 2) to explain the workings and functions of EXCOTRA which is developed to simulate world coal trade patterns.

Background of EXOCTRA

Market behaviors in world coal trade are very difficult to predict. For instance, in making coal purchasing decisions potential buyers in Korea will consider not only transportation and overall costs of coal but also some government policies with respect to "diversity of supply" and delicate political situations such as trade frictions with some foreign countries. The decision-makers in Korea may purchase high-priced and/or low quality coal as a means of easing trade friction or as a part of detente policy toward some communist countries.

Coal consumption in Korea is projected to dramatically increase from 14 million tons in 1986 to 51 million tons by the year 2001. All of this new coal demand will have to be imported, since virtually all available coal in Korea has been consumed (Kim, 1987).

While a total mass emission control system has not been adopted in Korea, a series of stricter environmental regulations have evolved since 1977. Ironically, however, it is expected that stricter environmental regulation will further discourage the use of coal only temporarily until clean coal technologies become economically viable (Kolstad, 1986). At the same time, interest in Korea's "quality of life" is expected to become a high priority over the next decade, promoting stricter environmental regulations. Thus, the level of clean coal technologies and environmental regulations are also important variables which determine future market patterns in Korea.

Due to many uncertain future events, it is difficult to determine relative strengths of coal from a particular country. In this context, decision-makers in both the public and private sectors involved in coal exports need an effective computer system which can help them predict possible outcomes of future events related to the coal marketing environment. Questions which are of particular interest to decision makers include:

1. What would be the impacts on global coal trade if Australian miners are on strike?

2. What would be the marketability of Illinois's high sulfur coal if clean coal technologies become economically viable in Korea?

3. What would be the impacts of opening of diplomatic relationships between Korea and China on the position of the U.S. coals?

EXCOTRA is developed to address these types of questions under different scenarios and assumptions. The primary objective of EXCOTRA is to design and implement a prototype simulation model using expert systems techniques that could serve as a tool for exploring various strategies for marketing coal. It is acknowledged that the report by Kim et al. (1988) to the Illinois Department of Energy and Natural Resources, which evaluates the marketing opportunities of Illinois steaming coal in Korea and Japan, provides the basis of building EXCOTRA.

Expert Systems and Simulation

In general, an expert system may be defined as a computer system that incorporates qualitative knowledge of human experts in a particular domain to perform functions similar to those normally performed by human experts. Expert systems is a branch of artificial intelligence which focuses on the development of advanced computer technology such as robotics, natural language, computer vision, and machine learning.

Since the late 1970s, expert systems have been applied to various fields which include chemistry, geology, medical, legal and military fields. Expert systems have also been researched for their applicability to planning problems (see Han and Kim, 1987 and Suh, Kim, and Kim, 1986 for example).

The knowledge representation characteristics of expert systems are one of the major factors which make expert systems a suitable tool for building the world coal trade simulation model. Expert systems can effectively represent the relations between objects involved in the simulation by means of production rule, frame, or network structure. In addition, McKenna (1980) views artificial intelligence and heuristic programming as one type of simulation, called "cognitive simulation," in that many simulation models attempt to match human problem-solving patterns and behavior. Human problem-solving involves mostly heuristics that can be explained as "rules of thumb." As discussed by many authors, expert systems have been effective in capturing heuristic knowledge of human decision-makers (Waterman, 1986 and Goodall, 1985). Gaschnig et al. (1981) even defines expert systems as "computer programs incorporating judgment, experience, rules of thumb, intuition and other expertise" of human experts. The linkage between expert systems and simulation models is the rationale of developing an expert system which simulates global coal trade patterns.

The heuristic approach of expert systems can be relied on when there is no analytical procedure for solving the problem. In the case of world coal trade, analytical techniques like linear programming may not be effective because the objective is not sufficiently well defined to make optimizing even meaningful. In fact, minimizing importing costs may be only one of the many concerns of most model builders for the coal trade, leaving important decision variables such as diversification of supply sources unaccounted.

When the objective is not clear, a heuristic method of expert systems can be used to find an acceptable solution rather than an optimal one. The following are examples of heuristics designed to project market share of major coal exporting countries in Korea:

- When coal miners in Poland are on strike, South Africa is likely to divert its coal export to Europe from Asia and the market share of American coal will increase in Korea.

- To be exported to Korea, the heating content (calorific value) of coal should be more than 6000 Kcal/kg when its sulfur content exceeds 2 percent.

- When clean coal technology is not feasible and sulfur content of coal exceeds 3 percent, there is no way to export coal to Korea.

The decision-makers in Korea may follow the rule of "at least 3 supply sources and less than 50 percent of total imports from one supplier" when the world coal market is in a normal situation (e.g. no major strikes and no trade sanctions).

The heuristics (or rules-of-thumb) illustrated above can be effectively encoded into the program using knowledge representation techniques of expert systems. For EXCOTRA, "IF premise, THEN conclusion" rules are used. One can easily imagine how effectively the heuristics shown above can be translated into IF-THEN rules.

Like any other simulation model, the expert system designed for simulation purposes enables decision-makers to experiment with different variables in order to answer "what if...?" questions by allowing decision-makers to develop their own scenarios regarding future events. In simulating real situations, expert systems may provide more realistic answers than other analytical models since expert systems attempt to follow the problem-solving behavior of human decision-makers by capturing their heuristics.

Structure and Function of EXCOTRA

EXCOTRA is an expert system which simulates world coal trade patterns based on the heuristic knowledge about the operation of the international coal markets. Because it is developed as a prototype system to determine the feasibility of building coal trade simulation models using expert systems, actual workings of the world coal market are very much simplified in EXCOTRA.

Components of EXCOTRA

EXCOTRA is an expert system coupled with a traditional database system and procedural programming techniques. By incorporating a traditional database system, EXCOTRA is designed to efficiently maintain a database for the information on the coal exporting countries such as coal quality specifications and price. Through the database management system, users can modify or update the EXCOTRA database without having to rewrite IF-THEN rules.

The coal quality and price data may be stored in a knowledge base using IF-THEN format. But it would be a very time-consuming and inefficient way of storing data. In fact, the coupling of the expert system and database management system has been a hot research topic in the expert systems field in an effort to overcome the limitations of traditional expert systems (Zobaidie, A. and J.B. Grimson, 1987; Brachman, Ronald J. and Hector J. Levesque, 1987; Jakobson, G., C. Lafond, and E. Nyberg, 1986).

The need for incorporating procedural programming techniques into EXCOTRA comes from the fact that it is difficult, though not impossible, to program predefined procedure in traditional expert systems. For instance, at the end of consultation, EXCOTRA needs to sort exporters using coal price and allocate import quantity in a way that minimizes importing costs. As shown in Figure 11.1, the component, Optimizer, is built using procedural programming to perform this type of task.

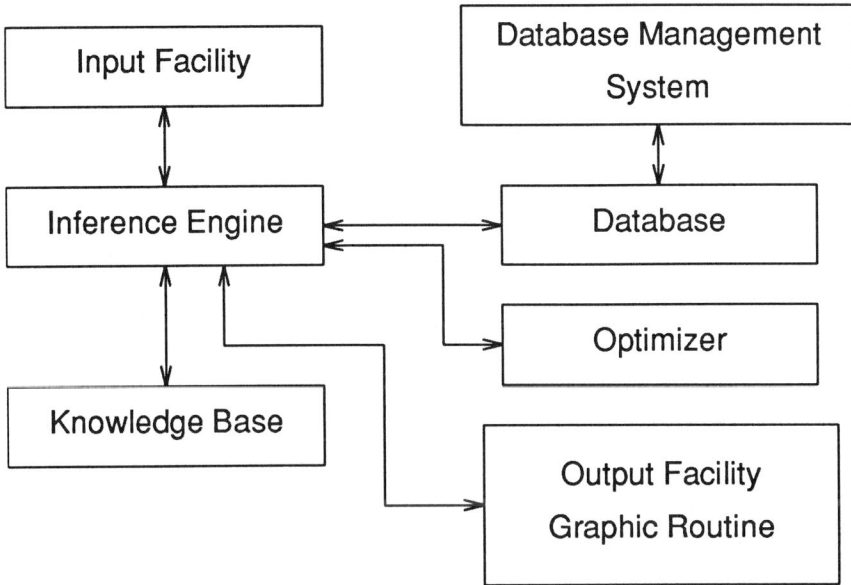

Figure 11.1. Components of EXCOTRA.

Other components of EXCOTRA include an Input Facility by which users develop their own scenarios regarding the future events affecting coal market situations, a Knowledge Base which maintains heuristic knowledge about the operation of international coal market and the behavior of coal buyers, and an Inference Engine which performs reasoning (backward chaining) to get a conclusion after consulting the Knowledge Base, Database, and Optimizer. The Output Facility summarizes consultation results in a report format and also plots a graph to show the projected market distribution. Figure 11.1 depicts these components and the relations between components.

Functions of EXCOTRA

The key function of EXCOTRA is to simulate future patterns of international steam coal trade. It attempts to model the decision-making process of coal buyers in Korea regarding coal purchasing decisions. The major processes involved in EXCOTRA may be summarized as follows:

1. Develop a scenario regarding world coal market situation. The major events with which users can develop scenarios include Miners' Strikes in Australia, Canada, South Africa, and Poland, Opening of Diplomatic Relationships Between Korea and China, Plunges (Surges) in the

Currency Values of Exporting Countries, Trade Sanctions Against South Africa, and Development of Clean Coal Technology. EXCOTRA helps users in describing these events by asking probabilities, year, duration and so on.

2. Calculate steam coal demand in Korea for the year specified by the user. To simplify the development process, EXCOTRA uses the demand projected by KEPCO (Korean Electric Power Company) for the year 1990, 1995, 2000, and 2010, and interpolates demand for other years. If time permitted, the possible variables to be used to project coal demand would include economic growth rate, industrial structure, and prices of substitute energy sources.

3. Examine existing supply conditions and develop new supply conditions based on the scenario developed in 1).

4. Develop coal quality requirements based on the scenario developed in 1) and check available supplies against required coal specifications.

5. Allocate coal demand obtained in 2) to qualified suppliers obtained in 4). The allocation procedures follow the rules in the knowledge base of EXCOTRA. The allocation rules deal with political preferences, government policies such as "diversity of supply," and coal price.

6. Plot a graph to show the projected market share in Korea.

7. Ask users whether they want to develop a different scenario. If yes, go to process 1), otherwise terminate consultation.

As illustrated above, the major function of EXCOTRA is to provide decision makers with an effective way of examining possible impacts of future events on the international coal market. EXCOTRA enables decision-makers to experiment with different variables in order to answer "what if" questions. For instance, the users of EXCOTRA can test the effects of introduction of clean coal technologies, plunges in the U.S. dollar value, or trade sanctions imposed on South Africa on the market position of their coal. The target users of EXCOTRA may be decision-makers of coal exporting countries who are responsible for developing marketing strategies to promote coal to overseas market.

Limitations

Because EXCOTRA is developed as a prototype system, it has some limitations on replicating real operations of the international coal market. These limitations are due to the limited number of variables included in EXCOTRA. The variables that may enhance the function of EXCOTRA include infrastructure conditions of importing as well as exporting countries such as port capacities and railroad conditions, and more realistic estimation of coal reserves of exporting countries.

The most critical problem faced in developing EXCOTRA is to devise the underlying relationships between variables. In the case of coal demand and oil price, the cross-elasticity between these two variables could be obtained from past data. For most variables, however, this type of relationship is difficult to devise. The integration of a machine learning program which discovers hidden knowledge (i.e. relationships between variables) from the database may enhance the reliability of EXCOTRA. After expanding the existing database of EXCO-TRA, commercial machine learning programs such as IXL: The Machine Learning System (Intelligence Ware, Inc., 1988) can be integrated into EXCOTRA.

Summary and Conclusions

EXCOTRA is a prototype expert system which is designed to provide decision-makers with an effective way of examining possible impacts of major future events on the international coal market. EXCOTRA may be regarded as a descriptive system rather than a normative one since it attempts to describe the market behavior under certain conditions rather than seek an optimum solution under those conditions.

To build an effective simulation model, EXCOTRA incorporates a database management system and procedural programming techniques into an expert system, overcoming limitations found in typical expert system structures. These limitations include inefficiency in handling large quantities of data and programming predefined procedures.

The target users of EXCOTRA may be the decision-makers of coal exporting countries who are responsible for developing marketing strategies to promote coal to overseas markets. EXCOTRA can be developed as a full scale expert system by incorporating more variables, such as substitute energy sources and new coal transportation technologies, and by eliminating some rigid assumptions which were necessary to shorten the development process of this prototype system.

References

Brachman, Ronald J. and Hector J. Levesque, 1987. "Tales from the Far Side of KRYP-TON: Lessons for Expert Database Systems from Knowledge Representation," *Proceedings from the First International Conference on Expert Database Systems.*

Gaschnig, J., R. Reboh and J. Reiter, 1981. "Development of a Knowledge Based Expert System for Water Resources Problems," SRI International, August.

Goodall, Alex, 1985. *The Guide to Expert Systems,* New Jersey: Learned Information, Inc.

Han, Sang-Yun and T. John Kim, 1987. "An Application of Expert System in Urban Planning: Site Selection and Analysis," *Planning Papers,* Department of Urban and Regional Planning, University of Illinois at Urbana-Champaign.

Intelligence Ware, Inc., 1988. *IXL: The Machine Learning System, User's Manual.*

Jakobson, G., C. Lafond, and E. Nyberg, 1986. "An Intelligent Database Assistant," *IEEE Expert,* Summer.

Kim, T. John, 1987. "Toward Developing Strategies for Promoting Illinois Coal to Korea," A Report to the Illinois Department of Natural Resources.

Kim, T. John, C.D. Kolstad, D.L. Fields, Sang-Yun Han, and Sunduck Suh, 1988. "Marketability of Illinois Coal to Korea and Japan," Report to the Illinois Department of Energy and Natural Resources.

Kim, T. John and Sunduck Suh, 1988. "Prospects for Illinois Steam Coal in the Pacific Export Market: An Overview," Transactions, *Society of Mining Engineers of AIME,* (282), pp. 1876-1882.

Klahr, Philip, John W. Ellis, William D. Giarla, Sanjai Narain, Edison M. Cesar, and Scott R. Turner, 1986. "TWIRL: Tactical Warfare in the ROSS Language," *Expert Systems: Techniques, Tools, and Applications,* edited by Philip Klahr and Donald A. Waterman, Reading, Massachusetts: Addison-Wesley Publishing Company.

Kolstad, Charles C., 1986. "Equity and Efficiency in Acid Deposition Regulation," *Working Paper,* Institute for Environmental Studies, University of Illinois at Urbana-Champaign.

McKenna, Christopher K., 1980. *Quantitative Methods For Public Decision Making,* New York: McGraw-Hill Book Company.

Suh, Sunduck, Moonja Park Kim, and T. John Kim., 1986. "An Expert System for Manufacturing Site Selection," *Planning Papers,* Department of Urban and Regional Planning, University of Illinois at Urbana-Champaign.

Waterman, Donald A., 1986. *A Guide to Expert Systems,* Massachusetts: Addison-Wesley Publishing Company.

Zobaidie, A. and J.B. Grimson., 1987. "Expert Systems and Database Systems: How Can They Serve Each Other?," *Expert Systems,* February.

CHAPTER 12

CGS-DSC: An Expert System for Construction Contract Claims[1]

Moonja Park Kim and Kimberley Adams

Among the most significant successes in the field of artificial intelligence has been the development of expert systems. These systems are designed to represent factual knowledge in specific areas of expertise and to provide the power of these systems has lead to a worldwide effort to apply this technology to various areas of expertise.

The expert system technology available today on microcomputers makes it possible to address a significant problem facing the construction industry: the need for expertise in construction claim analysis at the field level. Construction in the 1980's has become a very complicated industry, with many intertwined relationships and intense competition. There seems to be a great potential for applying the state-of-the-art knowledge in expert system technology to the practical areas of construction. One promising area is using an expert system to help minimize some of these problems by providing field personnel some guidance in handling legal issues related to potential claims.

The professionals in the legal field are also taking advantage of this new technology, as evidenced at the First International Conference on Artificial Intelligence and Law held May 27-29, 1987. Some law firms are even creating expert system groups to perform in-house research and development. For example, Watt, Tiedler, Killian and Hoffar, a law firm in Virginia, is developing a microcomputer expert system for claim identification and evaluation (Lester, 1987). The applications of this technology will assist lawyers in sorting out pertinent information for efficient discussion with the client.

---

1   The original version of this chapter is to be published in a forthcoming issue of *Construction Management and Economics*.

Researchers at the U.S. Army Construction Engineering Research Laboratory (USA-CERL) have been developing an expert system called Claims Guidance System (CGS) to provide claims analysis expertise at the field level of the Corps of Engineers. This system uses an expert system shell for IBM-compatible microcomputers, Personal Consultant Plus by Texas Instruments. The objectives of this system are (1) to provide an inexperienced project engineer with prelegal assistance in the analysis of potential claims from construction contracts and (2) to serve as a training device for new personnel in field offices, familiarizing them with the related legal issues.

The first module of the Claims Guidance System (CGS-DSC) will guide project engineers in analyzing "Differing Site Condition" (DSC) claims. Unknown subsurface conditions or latent physical conditions at the work site represent a very significant risk inherent in many construction contracts. The DSC contract clause represents an effort by the U.S. Government to reduce the risk to the construction contractors of such unknown or unanticipated conditions. This clause allows contractors to submit their bids based on reasonably foreseeable conditions, without contingencies to cover the unexpected or unusual. In return, the bidder is assured that in the event conditions prove different than should have been anticipated, an equitable adjustment will be made in the contract price and/or duration. Without this clause, the contractors' only alternative would be to raise the bid price in order to meet the requirements, is covering the cost of coping with possible subsurface difficulties, which in fact may not have occurred during subsequent performance of the contract. As a result, the Government paid more than the actual work was reasonably worth.

Studies by Mogren (1986) and Diekmann & Nelson (1986) have shown that DSC claims are one of the most frequent and costly reasons for changes in U.S. Army Corps of Engineers construction contracts. Corps field engineers who are faced with such claims need to understand the legal issues involved so that they can supply the proper information to legal counsel and avoid lengthy litigations caused by incorrect decisions. Personnel who are unfamiliar with this process must rely on experienced engineers for help in analyzing a claim. The expert system for Claims Guidance is to provide the expertise of the experienced engineers in dealing with construction contract claims.

Specifically, an expert system for analyzing potential claims insures that a rigorous evaluation is performed consistently. It provides a written document of the claim analysis for future reference, which is especially useful if the claim must be defended. In addition, repeated use of the expert system sharpens the field engineers' claims evaluation skills which will help them identify potential claims sooner, avoid conflicts if possible, and support their position with adequate documentation.

This paper describes the expert knowledge acquisition process and the development of the CGS-DSC and a sample consultation session.

Acquisition of Expert Knowledge

Diekmann & Kruppenbacher (1984) have demonstrated that there is significant potential for applying artificial intelligence to claims analysis. They identified the need for more development work in this area to make this technology a viable tool for construction professionals in claims analysis. Following their suggestion and taking advantage of their work on knowledge acquisition, the DSC clause was selected as the first module of CGS for actual use in the Corps field offices.

The expert knowledge acquisition process for the CGS-DSC started with an analysis of the Differing Site Condition Clause (FAR 52.236-2) used by the U.S. Government in its contracts which cites:

a. The Contractor shall promptly, and before the conditions are disturbed, give a written notice to the Contracting Officer of (1) subsurface or latent physical conditions at the site which differ materially from those indicated in the contract or (2) unknown physical conditions at the site, of an unusual nature, which differ materially from those ordinarily encountered and generally recognized as inherent in work of the character provided for in the contract.

b. The Contracting Officer shall investigate the site conditions promptly after receiving the notice. If the conditions do materially so differ and cause an increase or decrease in the Contractor's cost of, or the time required for, performing any part of the work under this contract, whether or not changed as a result of the conditions, an equitable adjustment shall be made under this clause and the contract modified in writing accordingly.

c. No request by the Contractor for equitable adjustment to the contract under this clause shall be allowed, unless the Contractor has given the written notice required; provided, that the time prescribed in (a) above for giving written notice may be extended by the Contracting Officer.

d. No request by the Contractor for an equitable adjustment to the contract for differing site conditions shall be allowed if made after final payment under this contract.

From the above clause, the following important issues in dealing with DSC claims are identified:

1. Final payment: A Contractor that has accepted the final payment is not allowed to file a claim, as described in (d) above.

2. Notice Requirements: The Contractor must give proper notice of the differing site condition in order to maintain the possibility of entitlement as described in (a) and (c) above. Key aspects here are promptness, the receipt of the claim notice by the responsible Government personnel, and the written form of the notice.

3. Prejudice to Government: If a failure to give written notice has not resulted in any prejudice to the Government, the Contractor's right to relief on a valid claim will not be barred. If the Government's interests have been impaired by the Contractor's failure to give proper notice of the differing site conditions, it is unlikely that the Contractor will be entitled to compensation. To check if the Government was prejudiced, it is necessary to check if the Government would have directed the same actions had it received the appropriate notice.

4. Government Action: The contracting officer must look into the problem of the site condition as soon as he receives the notice in order to discuss the problem with the contractor and to direct the actions to be taken by the contractor for the problem, as described in (b) above.

5. Contract Provision (Type I or Type II): The DSC clause gives two avenues of recovery for the contractor, depending on the contract provisions that exist in each particular case. A Type I case is characterized in (1) of the above clause. The contractor can be entitled to compensation if the actual conditions differ materially from the conditions explicitly or implicitly mentioned in the Contract. A Type II case is characterized in (2) of the above clause. The contractor can also be entitled to a compensation if there is no specific indication of the condition in the contract documents, and if it can be demonstrated that by following the standard construction practice the Contractor could not expect, nor detect, the presence of differing site conditions prior to bid time.

6. Acceptable and prudent: When there was no indication of condition in the contract document, the contractor can make assumptions that are generally accepted practice in the construction industry. These assumptions must be the same as what a prudent contractor normally would use (assume usual and known conditions).

In addition to the analysis of the DSC clause, the work of Diekmann and Kruppenbacher (1984) was considered in the process of knowledge acquisition. The logic diagram of Kruppenbacher's (1984) study was reviewed by an experienced Corps field engineer and was revised and simplified to fit the Corps office environment. The questions used in Kruppenbacher's system include many legal terms that could confuse the field engineers; therefore, questions for the CGS-DSC were changed to be easily understandable by the Corps field office personnel. Using the revised logic diagram and questions, rules were developed to create a test version of the CGS-DSC.

A steering committee was formed to review the test version and to evaluate it for validity and completeness. The committee consisted of six experts: two experienced legal counsels from the Corps headquarters and four engineers with many years of experience in construction contract management within the Corps. The committee suggested many enhancements and necessary corrections to the logic diagram and identified important additional claims issues including reliance on contract information, superior knowledge, nature of condition (Act of God), site inspection, material difference, exculpatory language, and anticipation: usual and known.

Two legal case retrieval systems, Lexis and Weslaw, were used to select appropriate cases which represent some of the issues listed above. Twenty-three relevant cases were retrieved. These cases were examined carefully in terms of the important issues and of the rationale for the decision of the Board of Contract Appeals and/or the Court of Claims.

Short descriptions of four of these cases are included here to explain how different cases represented different issues in the CGS-DSC. The first case, C.H. Leavell v. Eng BCA (No. 3492, 1975), added the issue of reliance. In this case the Contractor's claim for a DSC equitable adjustment was denied because the Contractor mistakenly relied on inconsistent contract information when more detailed information was available. The Contractor would have discovered the information if he had followed the leads in the contract itself. The Contractor also failed to make his DSC claim until one year after the work was completed. Since the Contractor failed to meet the notice requirement, the Government was prejudiced and could not defend the case adequately. This contract was to construct five buildings at Lackland Air Force Base. The contract's borings indicated that the subsurface soil was practically impervious. Some of the drawing's symbols were unclear but the Contractor assumed that the symbols represented impervious soils and that water would not enter the holes after drilling. Unfortunately, the soil was pervious and extensive casing was necessary for many of the drilled piers.

The second case, the Portable Rock v. U.S. (Court of Claims, 1984), added the issue of superior knowledge. The Contractor sought an equitable adjustment for a Type I DSC condition when he found subsurface water conditions, not specifically mentioned in the contract, which substantially increased the costs of performing the contract. The Contractor alleged that the Government had knowledge of this condition but did not reveal this information to the bidders. The Contractor's claim for an equitable adjustment was denied. The Court held that the Government did not withhold information from the Contractor. Furthermore, the contract interpretation must be reasonable, and it was not in the case. The requirements in the contract, such as the reinforced subgrade, were a clear indication of unstable soil conditions. This should have alerted a knowledgeable contractor to the presence of ground water and wet ground conditions. This Contractor had superior knowledge because he had encountered wet conditions in this area when constructing another road, so he was aware of the conditions.

The third case, Turnkey Enterprises, Inc. v. U.S. (Court of Claims, 1979), added the issue of "Act of God" to CGS. The contract was for the repair of seven damaged sites on the Mad River Road in the Six Rivers National Forest in northern California. The Contractor claimed both Type I and Type II DSC conditions when a drought caused his costs to increase. The Court denied both Type I and Type II claim for an equitable adjustment, because a drought is an

"Act of God" and not any fault of the Government. And also the contract documents do not indicate anything about the alleged changed condition. There was nothing in the contract on which the Contractor could claim he relied upon. There was no guarantee that water would be in the river at all times. Generally, the Government does not assume an obligation to compensate a contractor for additional costs or losses incurred as a result of solely weather conditions.

The fourth case, Parker Construction Co. v. U.S. (Court of Claims, 1970), added the issue of reasonable site inspection. The Contractor entered into a unit price contract to build 2.366 miles of graded road in Washington State Mt. Baker National Forest. The Contractor claimed a Type II DSC because of hard rock encountered that increased his drilling costs. The Government claimed that the Contractor did not conduct a prudent site inspection. The Contractor assumed that the brownish cleavage faces of the rock indicated that the rock was softer and easier to drill than solid gray rock. He assumed this even though there were outcrops of hard rock apparent upon visual examination. Furthermore, the Contractor failed to prove his Type II DSC claim. The Contractor had to prove that he encountered something materially different from the "known" and "usual." The Court denied the Contractor's claim for an equitable adjustment.

Development of the CGS-DSC

The logic diagram for the CGS-DSC was developed based on important issues identified from the analysis of cases in combination with the logic diagram of Kruppenbacher (1984). The logic diagram and listing of 23 retrieved case briefs are found in the reference manual of CGS.

Based on the logic diagram, rules for the CGS-DSC knowledge-base were written using Personal Consultant Plus. There are about 300 rules and 95 questions that require the user's input in this knowledge base. The user is not asked all of 95 questions in every consultation; only a portion of them will be asked depending on the answers the user enters.

In selecting an expert system shell for the development of the system there were some limitations: (1) consultation should be available on a IBM-compatible computer with 640K memory, (2) cost of developing and delivering system should be minimized. Personal Consultant Plus was selected to meet these limitations and to provide a user friendly interface.

As we added rules to the knowledge-base, it was necessary to add 1.5 megabytes of memory for the development stage. However, every effort was made to make the delivery system run with 640K memory, by creating disk-resident text files for help screens and case summaries. As shown on Figure 12.1, the CGS-DSC delivery system environment was created by displaying text files for help screens, by presenting PC Paint graphic images for progress checks, and by running dBASE III Plus to search for the appropriate case name of the text file to be displayed. In order to create a useful report for the users, dBASE III Plus was used to organize input entered by the user and to provide a documentation of facts. An example of a help screen and a case summary are included for the sample consultation discussed below.

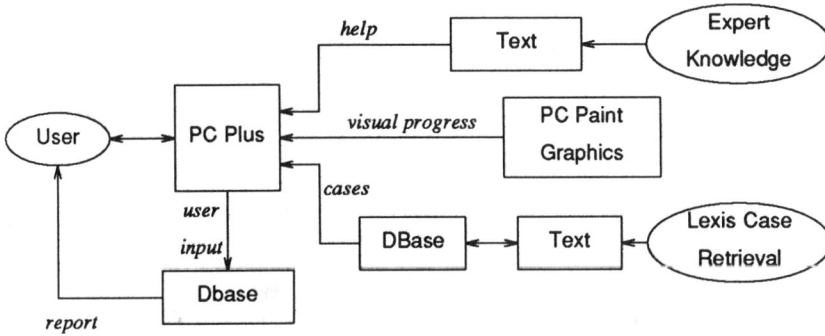

Figure 12.1. CGS-DSC environment.

The field test version was delivered to two test sites with a proper training. After the field test, the system will be distributed to a number of field offices for regular uses.

A Sample Consultation

One case from the 23 retrieved cases was chosen to be used as an example for explaining the use of the CGS-DSC: Hallmark Electrical Contractor Inc., v. ASBCA No. 32595 (April 27, 1987). A short description of this case is as follows:

> The Contractor was awarded a contract to replace an overhead electrical distribution system at the Belton Lake Recreational Area near Fort Hood, Texas. The Contractor was required to dig holes approximately six feet deep and place wooden poles in them. The Contractor estimated that it would take about an hour to dig each hole because he had worked on similar terrain about ten miles away from the present site at Fort Hood. Unfortunately, the Contractor encountered problems in digging and could not use the standard combination dirt/rock bit. The Contractor encountered "blue rock" which he had never encountered and which was much harder than the more prevalent "butter rock." The Contractor submitted a claim under both Type I and Type II DSC. The Contractor was denied recovery under Type II DSC

because the condition encountered did not differ materially from the "usual" and "known." Contractor was not entitled to recover under Type I DSC either, because the contract did not indicate conditions different from what the Contractor encountered during performance.

A Final Payment release was not signed by the Contractor. The Contractor complied fully with the notice requirement by filing a timely claim with the Contracting Officer for the adjustment. It was denied in its entirety by the Contracting Officer. The contractor anticipated butter rock and encountered blue rock and argued that blue rock is unusual for the area where he worked. However, an expert witness testified that "blue rock" is usual and known for the area. Therefore, the Contractor was not prudent to anticipate butter rock when submitting his bid. The Contractor did not conduct a reasonable site inspection because he assumed that the absence of a "rock table" is an "indication" that rock would not be encountered. It was apparent during a site visit because there were rock outcroppings all over the site, but he did not take this into consideration in his bid.

Rationale For Denying An Equitable Adjustment: The Contractor failed to prove that he encountered something materially different from the "usual." It was true that the blue rock was significantly harder than the butter rock. This in itself did not establish a Type II DSC. The Contractor had an expert testify, but the expert did not establish that the hard material was geologically unusual in the area. In fact, he established just the opposite. Since the Contractor encountered material within the range of hardness usually encountered, the Contractor was denied his Type II DSC.

This case will be used for demonstrating a sample consultation of the CGS-DSC. The user starts the consultation by typing "CONSULT CLAIMS". Then the title and the objectives screen will be displayed. The objective screen is shown below:

```
CLAIMS GUIDANCE SYSTEM
This Claims Guidance System is developed
to provide:

  1) basic awareness of the issues
     surrounding the DSC clause,
  2) measuring device to ascertain
     contractor's chance of DSC entitlement,
  3) documentation in the event the dispute
     comes to trial.
 * End - RETURN/ENTER to continue
```

Next, the user is asked to type in relevant information for documentation, such as the name of the contractor, contract number, so on.

```
Q: Please enter the name of the contractor.....
        Hallmark Electrical Contractor's, Inc.
Q: Please enter the contract number.....
        DAKF48-85-C-0052
```

```
Q: Please provide the description of the contract.....
       Overhead Electrical Distribution System
Q: Please enter your name (the user).....
       John Smith
Q: Please give a brief statement of the contractor's assertion.....
   The contractor could not dig holes using the standard
   combination dirt/rock bit in the one hour anticipated.  It
   took as much as a day or more to dig each hole where
   'blue rock' was encountered.  It took about 45 days to
   drill all the holes rather than the ten days the contractor
   had anticipated.
```

For all questions displayed below, the answers are shown in the parenthesis ( ). However, in actual consultation, the user will move the cursor:

```
Q: Has the contractor signed the final payment release without
   condition ?. . . . . . . . . . . . . . . . . . . . . (NO)
Q: Was the final payment sent to the contractor ? . . . . (NO)
```

The next step is to find if the contractor had complied fully with notice requirements. In order to check if appropriate notice was given, the following questions will be asked:

```
Q: Did the contractor inform the government about the problem prior
   to asserting a differing site condition ? . . . . . . (YES)
Q: How did the contractor inform the Corps of the problem?. . . . . . .
   . . . . . . . . . . . . . . . . . . . . . . . . . . . (WRITTEN)
Q: Please enter the name of the person that received  the information
   first.....
       Mr. James Kemp
Q: Please enter the date the information was received.....
       JUNE 27, 1985
Q: Who was the first person or group to receive information regarding
   the claim ? . . . (RCO/COR OR AUTHORIZED REPRESENTATIVE)
```

At any point before answering a question, the user can invoke the WHY option requesting to explain why the system is prompting for this information. The why screen for the above question is shown as an example:

```
WHY

Who was the first person to receive the
notice is needed to determine if responsible
person did receive the notice.

RULE 008
In order to check if the contractor complied
fully with the requirements for the notice
the following information is necessary:

(1)  if responsible personnel received the
     notice and the date received,
(2)  if the form of notice was adequate
     (written or oral)
(3)  if the notice was prompt.

     ** End - RETURN/ENTER to continue

Q: When did the contractor first inform the government about the
   problem ? . . . . . . . . . (BEFORE STARTING WORK ON PROBLEM)
```

From the answers shown above, a conclusion can be made that the contractor complied fully with notice requirements by informing the responsible personnel in a written form before starting work on the problem. Now we need to check if the government's action was adequate with the following questions:

```
Q: Did the government investigate the asserted Differing Site
   Condition ? . . . . . . . . . . . . . . . . . . (YES)
Q: Did the government's investigation confirm the contractor's
   asserted DSC ? . . . . . . . . . . . . . . . . (NO)
Q: Please summarize the investigation made by the Government. . .
      Government decided not to presurvey  and represent the
      subsurface conditions.  The contractor must demonstrate
      that he has encountered something materially  different
      from the 'known' and 'usual'. Even though the government
      found that  blue rock was harder than the softer material
      that the contractor hoped for, this is not unusual around
      this area.
Q: Did the government provide the contractor with directions
   concerning the asserted DSC ? . . . . . . . . . (NO)
```

At various points a graphic display of progress through the system is available. An example of this graphic display after checking the Government action is shown in Figure 12.2. This graphic display is an over-simplified version of the CGS-DSC logic diagram. It informs the user how many of the issues were covered through the system at various stages.

The next step in the CGS is to check the nature of the problem, material difference, contract provisions, usual and known anticipation issues.

```
Q: Is the asserted DSC directly related to the physical conditions at
   the site ?. . . . . . . . . . . . . . . . . . . . . (YES)

Q: Did the asserted DSC exist before the contract award ? (YES)

Q: Does the physical condition substantially differ from the
   indicated/anticipated condition ?. . . . . . . . . (PROBABLY YES)

Q: Please describe why you believe that the condition differs
   substantially. . .
      Because the blue rocks the contractor encountered are much
      harder than what the contractor hoped to encounter, and it
      took much longer to drill holes which resulted in substantial
      increase of cost.

Q: Did the asserted DSC increase the contractor's costs/duration ?
   . . . . . . . . . . . . . . . . . . . . . . . . . . (YES)

Q: Did the contractor utilize acceptable practices or standards of
   workmanship to alleviate the DSC ?  . . . . . . . . (YES)

Q: Are there contract indications or notes on drawings that pertain to
   this situation ?. . . . . . . . . . . . . . . . . . (NO)
```

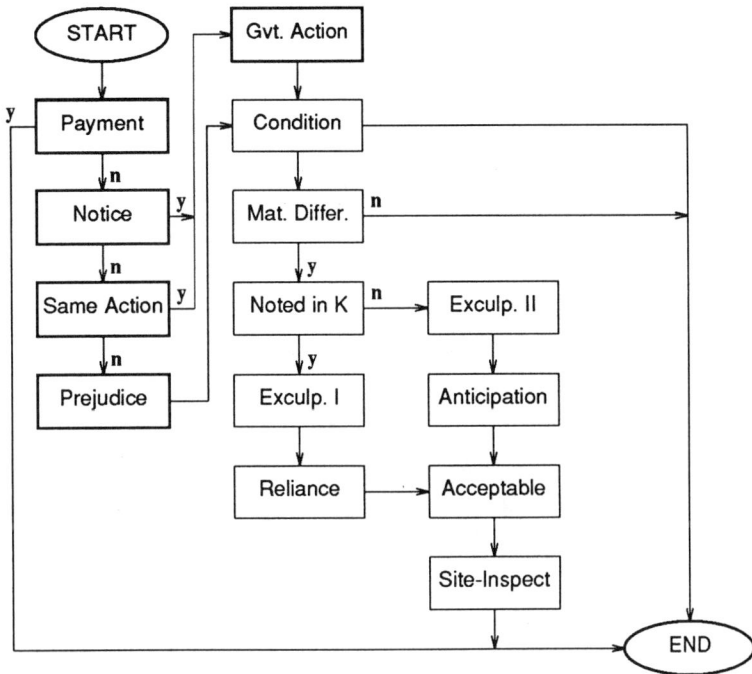

Figure 12.2. Simplified logic diagram.

Q: Please explain in detail why you believe that the contract does not
   either implicitly or explicitly make reference to the alleged differing
   site condition claim. . . . .
       The government has elected not to presurvey and represent the
       subsurface conditions, so there was no indication about the
       condition in the contract.

Q: Please explain the known and usual conditions in the area. . . . . .
       Blue rocks can be found scattered around this area and the
       hardness of these blue rocks varies a great deal.

Q: Please explain the conditions that were encountered in the area. . . . .
       In one third of the holes, hard blue rocks were encountered so
       that the contractor spent a day or more to drill each hole.

Q: Is there a clause that limits the government responsibility and
   liability by stating that indications on the contract are only a
   representation of conditions and should not be a basis for differing
   site condition ?. . . . . (YES)
Q: Is this clause extremely specific to the condition encountered ?. . .
   . . . . (NO)
Q: Did the contractor act prudently in making assumptions about the
   site condition ? . . . .   (DEFINITELY NO)

At any time during the consultation before answering a question, the user can display a help screen for explanation of the question, as additional clarification. The help screen for the last question above is:

```
PRUDENT CONTRACTOR:

If most contractors with similar experience and
knowledge, and under the same circumstances,
would agree with the contractor in question,
the contractor is said to be a prudent one.

Prudent Contractor would:

(1)   Utilize his past experience and knowledge
      as a basis for all his judgments and
      assumptions.
(2)   Act as most of his peers would given the
      same set of circumstances.
(3)   Submit contract bid based on fair reading
      of representations indicated in government
      contract and adequate site inspection.
(4)   Make assumptions utilizing his past
      experience and knowledge when conducting
      a site inspection.
(5)   Make assumptions utilizing his past
      experience and knowledge when reading
      the contract drawings, specs., and,
      therefore, not rely on obvious flaws
      noted in the contract.

Q: Please explain why you think that the contractor was not prudent....

      He should have checked what is usual around this area, and he
      should have noticed from the site visit of rock outcroppings all
      over the site.

Q: Should the contractor have known about the DSC prior to bidding?
   . . . . . . . . . . . . . . . . . . . . . .    (DEFINITELY YES)

Q: Is there evidence that the contractor knew of the DSC prior to bid?
   . . . . . . . . . . . . . . . . . . . . . . . . .  (NO)

Q: Did the government have current knowledge of the DSC prior to bid?
   . . . . . . . . . . . . . . . . . . . . . . . . .  (NO)

Q: Would a reasonable site inspection have provided information to
   detect the DSC?. . . . . . . . . . . . . . . . . .   (YES)

Q: Did the contractor conduct a reasonable site inspection prior to bid?
   . . . . . . . . . . . . . . . . . . . . .  (DEFINITELY NO)

Q: Did any act of the government hinder the site inspection?
   . . . . . . . . . . . . . . . . . . . . . . . . . .  (NO)
```

This is the end of consultation for this case. Next, the conclusion screen will be displayed as shown below:

```
CONCLUSIONS

Possible conclusions for entitlement are the
following:
    1. VERY-POOR-CHANCE
    2. POOR-CHANCE
    3. DIFFICULT-TO-DECIDE
    4. FAIR-CHANCE
    5. GOOD-CHANCE
    6. EXCELLENT-CHANCE

The chance of entitlement for this contractor
is as follows:  VERY-POOR-CHANCE
```

After displaying the conclusion reached by the CGS, the system will search for a case similar to the situation described by the user. It may take a few minutes for the system to perform the dBASE III commands and to find the appropriate case. The user will see the dBASE III logo and license agreement statement, then the short summary of the retrieved case from the cases database would be displayed. For this consultation, a one-screen summary of the Hallmark case will be displayed.

Next, the list of the pertinent factors in reaching the conclusion is displayed on the screen as shown below:

```
The list of pertinent factors in reaching the
conclusion is as follows:

    CONTRACTOR EXPECTATIONS ISSUE
    SITE-INSPECTION ISSUE
```

After this, the user needs to save all his answers in a file in order to review it later. The answers and information which the user has supplied during the consultation can be compiled into a report so that the information can be easily and clearly presented in a fact sheet. This report generation is performed by dBASE III because PC Plus does not provide adequate reports. A sample report is included in the CGS users manual.

Future Research

Even though USA-CERL attempted to avoid legal terminology in writing questions for users (mainly engineers) the CGS still requires some legal judgment as input. For example, the user has to make a legal judgment to characterize the difference as "material" or "substantial" when answering the question "Does the physical condition substantially differ from the indicated condition?" It would be desirable to include more expert knowledge of lawyers on how to make a legal judgment on "materiality" and other topics. It seems that work on this area may be available in the near future if we look into the research and development being done in the law profession.

Many law professionals are actively exploring expert system technology, as evidenced at the first International Conference on Artificial Intelligence and Law held May 27-29, 1987. Waterman, Paul and Peterson (1986) reviewed existing expert systems for legal decision-making potential and indicated that they expect more applications in the following areas: organizing case information, estimating case value and establishing strategies for negotiation, monitoring legal data bases to find changes in the law, interpreting the law in the context of a problem, and producing legal documents.

Another approach was presented in Victor's (1984) article, "How Much is a Case Worth," which demonstrated how a collection of decision trees, subjective probability assessments and calculations can be used in evaluating claims; this application helps trial lawyers assess the monetary worth of alternative courses of action.

Thus, it seems that a great amount of research and development is expected in the near future. We may take advantage of this interest in the legal field and include more legal expertise in improving the CGS to include other types of construction contract claims. For example, including construction delay claims would involve integrating scheduling and network analysis with legal evaluation of claims. Design deficiency claims would involve integrating CADD systems with claims evaluation to examine drawings for deficiencies. These areas are challenging and hold potential benefit for the construction community.

Acknowledgement. The authors would like to express their gratitude to Vida Florez, student at Vermont Law School, South Royalton, VT 05065, for retrieving and reviewing appropriate cases for this paper.

References

Corps of Engineers, Department of the Army, 1986. *Contract Clauses - Construction - Inside the U.S.*

Diekmann, J. E., and T.A. Kruppenbacher, 1984. "Claims Analysis and Computer Reasoning." *Journal of Construction Engineering and Management,* Vol. 110, No. 4, ASCE, 391-408.

Diekmann, J. E., and M.C. Nelson, 1985. "Construction Claims: Frequencies and Severity," *Journal of Construction Engineering and Management,* Vol. 111, No. 1, ASCE, 74-81.

Kruppenbacher, T. A., 1984. "The Application of Artificial Intelligence to Contract Management." *USA-CERL Technical Manuscript P-166* (U.S. Army Construction Engineering Research Laboratory).

Lester, J. L., 1987. "Lawyer on a Microchip." *Civil Engineering Magazine,* ASCE, 68-69.

Mogren E.T., 1986. "The Causes and Costs of Modification to Military Construction," *Master's Thesis,* U.S. Army Command and General Staff College.

Victor, M. B., 1984. "How much is a Case Worth ? - Putting Your Intuition to Work to Evaluate the Unique Lawsuit," *Trial.*

Waterman, D. A., Paul, J., and Peterson, M., 1986. "Expert Systems for Legal Decision Making." *Expert Systems - The International Journal of Knowledge Engineering,* Vol. 3, No. 4, Learned Information Ltd., Oxford, 212-226.

CHAPTER 13

MEDIATOR: An Expert System to Facilitate Environmental Dispute Resolution

Yi-Chin Lee and Lyna L. Wiggins

As society progresses, people enjoy societal and technological well being from economic development, but also suffer from the environmental externalities caused by development. In the last two decades awareness of the negative consequences of economic development has escalated to such an extent that the public has become increasingly involved in conflicts over development issues. This increased concern about environmental impacts has provided a foundation for the prevailing environmental movement.

The increased environmental consciousness has many roots. Yet these roots all share one important element: They are all grounded in conflict. Public opinion polls show increased general concern with protecting natural resources but also reveal that views are sharply divided on specific issues such as balancing the need for national energy independence against protection of the wilderness.

Disputants have become accustomed to resorting to lawsuits to resolve environmental disputes. But lawsuits have disadvantages as resolution techniques, not the least of which are the time, effort, and money that the involved parties need to invest. As the litigious approaches to conflict resolution have become more and more costly and inefficient, a new body of theory and practice, based on negotiation in the presence of a neutral mediator or facilitator has developed (O'Connor, 1978; Mernitz, 1980; Lake, 1980; Kolb, 1983; Bacow and Wheeler, 1984; Folberg and Taylor 1984; Bercovitch, 1984; Moore, 1986; Harashina, 1988). From numerous case studies in the past decades, researchers find that this mediation approach for dispute resolution has advantages (Mernitz, 1980; Bacow and Wheeler, 1984) of: (1) being as speedy as the parties wish it to be; (2) being less expensive than litigation; (3) having a more systematic approach relying upon some principles to reach a final consensus; (4) being, in general, a voluntary function; and (5) making it more likely that substantive issues will be addressed.

Mediation has become more accepted as a viable alternative for environmental dispute resolution (Susskind, Bacow and Wheeler, 1983; Susskind, 1987; Bingham, 1986; Raiffa, 1982; Fisher and Ury, 1981; Buckle and Thomas-Buckle, 1986; Burgess and Smith, 1983; Cormick, 1973; Cormick and McCarthy, 1974). Although there is still confusion over the definition of the terms and the boundaries for analysis, it is evident that clarification is on its way. While the agendas of individual mediators, facilitators, and other professionals in dispute resolution may diverge, a set of common issues is clear. Practitioners feel the need to be able to explain to themselves and others why they operate the way they do as dispute resolvers. Researchers devote effort to conceptualizing a systematic approach to environmental dispute resolution practice. Such efforts are organized to explore the intersection of theory and practice, to formalize the rules-of-thumb that are effective in practice, and to accumulate the insights that seem to apply across contexts.

One possible direction of research in environmental dispute resolution is to consider the applicability of new research in the area of expert systems. What contribution could these new technologies make to environmental conflict management?

Expert Systems

Feigenbaum defined expert systems as (Feigenbaum, 1985):

... an intelligent computer program that uses knowledge and inference procedures to solve problems that are difficult enough to require significant human expertise for their solution. Knowledge necessary to perform at such a level, plus the inference procedures used, can be thought of as a model of the expertise of the best practitioners of the field.

The knowledge of an expert system consists of facts and heuristics. The facts constitute a body of information that is widely shared, publicly available, and generally agreed upon by experts in a field. The heuristics are more private, little-discussed rules of good judgment (rules of plausible reasoning, rules of good guessing) that characterize expert-level decision making in the field. The performance level of an expert system is primarily a function of the size and the quality of the knowledge base it possesses.

Essentially, expert systems should possess intelligence that can achieve high levels of performance in hard areas. Classic examples like MYCIN (an expert system that aids physicians in the diagnosis and treatment of meningitis and bacteremia infections, developed at Stanford in the 1970's) and PROSPECTOR (an expert system that provides consultation to geologists in the early stage of investigating a site for ore-grade deposits, developed at SRI International in 1981) are all successful from this perspective. These systems capture the expert knowledge and experience that are required to solve problems in their respective areas. These early system development efforts suggest that, as long as knowledge and experience can be articulated into a systematic structure, the problem may fit into an expert system category. This led us to think of research into the articulation of the knowledge and experience that a mediator uses to resolve contradictions encountered in an environmental dispute, and of putting this domain expertise into an expert system.

In practice, due to the complex nature of human intelligence and problem solving abilities, today's expert system technology, which is also called knowledge engineering, can best handle problem domains that possess the following eleven characteristics (Waterman, 1986): (1) task does not require common sense; (2) task requires only cognitive skills; (3) experts can articulate their methods; (4) genuine experts exist; (5) experts agree on solutions; (6) task is not too difficult; (7) task is not too easy; (8) task is not poorly understood; (9) task requires symbol manipulation; (10) task requires heuristic solutions; and (11) task is of manageable size.

Many of the existing expert systems embrace these characteristics. The challenge we are facing today, when trying to develop an environmental dispute mediation expert system is that the task is still not clearly understood. Many experts have difficulties articulating the rules-of-thumb they use to tackle various situations. If we were to formalize portions of this environmental dispute mediation function into a systematic structure, putting it into an expert system would be the ultimate test. It was also clear that, in considering the 11 criteria for workable problem domains discussed above, to successfully implement an operational system we would need to narrow the scope of the problem domain.

Developing the MEDIATOR Expert System

We began the system design of MEDIATOR with the goal of developing a real-time meeting aid for a mediator involved in an environmental dispute resolution. The research process consisted of: (1) reviewing the literature and interviewing an experienced mediator to formalize the environmental dispute mediation task into a systematic structure; (2) formalizing the environmental dispute mediation task by breaking the structure down into components that were easier to tackle, selecting one of the components as a prototype problem, and developing a case example; (3) building a computer program that incorporates the idea of a component of the mediation process and the knowledge of an experienced mediator; and (4) validating the system's performance against that of the domain expert. These four research steps are described in more detail below.

Finding an articulate expert. One of the most important criteria for expert system development is the availability of a committed and articulate source expert. We were fortunate to have the continuing assistance of an experienced mediator in this research. Our domain expert has worked for several decades as a facilitator of numerous public policy controversies. Since he was interested in the project, and had some experience with using computer tools during meetings, he provided excellent, informed, and timely guidance to us throughout the research process.

Formalizing the environmental dispute mediation task. To limit the scope of the problem, and to define a task that has been completed by many mediators, we decided to focus on a facility siting problem. Siting is a process that involves many stages, and various siting methods are available. One of the common approaches used in a siting exercise is to select evaluation factors, assign a weight to each siting factor, and, for each site, assign a scored value to each siting factor. How, and if, a weighted sum of scores should then be constructed is a matter of some debate in the planning literature (Hopkins, 1977). Nevertheless, such a procedure is commonly used in siting practice.

There are various approaches used in practice to develop weights for siting factors. One procedure involves citizen groups and other participants voting, another involves a *weighting expert* who assigns weights to each siting factor (ERCDC, 1978; Keeney and Kirkwood, 1975); and another is a decision analysis approach proposed by Ralph Keeney (1980). Keeney uses decision analysis to determine a participant's opinion on weight assignments. This decision analysis approach has generally been used with only one "super decision maker" to determine weights.

A new approach is suggested in this research. Participants from multiple concerned parties sit down with a mediator and an expert system aid to work out a consensus on these siting factor weights. Such a task possesses many of the characteristics of the resolution of other components of environmental disputes; and, most of all, it is of manageable size. In completing the task of developing weights for siting factors, larger issues of participant identification, meeting management, etc. will have to be addressed.

The problem formalization outlined above considerably simplified our design task. However, designing a prototype system to encompass all manner of facilities (e.g., power plants, affordable housing, parks), at all sites, and in all regulatory environments was still too broad. We therefore selected a representative siting problem by narrowing to a specific case of a municipal-solid-waste (MSW) energy facility in the state of California. This allowed us to use actual case information from a real facility siting exercise as base data.

Developing the prototype system. In the process of constructing MEDIA-TOR, we made use of the concept of reverse knowledge engineering (Sowa, 1988) to clarify the task and sort out the important features of the system. The basic idea of reverse knowledge engineering is to first conceptualize a working system, then to work backward to figure out what are the required features. This concept somewhat coincides with the approach introduced by Shachter and Heckerman (1987).

The working system was envisioned as an expert system that manages a multiparty-negotiation meeting environment. The meeting's purpose is to determine an agreed-upon weighting scheme for site selection factors among interested parties. The system's basic configuration emulates a taxonomy of the mediation function, and includes: background knowledge acquisition, participant identification, siting factor determination, preference quantification, data analysis, conflict detection, conflict interpretation, and conflict resolution. This configuration is illustrated in Figure 13.1. In this configuration, the first four modules comprise pre-meeting fact finding tasks; while the latter four perform the tasks of joint fact finding, or "out in the field negotiation (Susskind, 1987)." Analytical calculations and inferences within MEDIATOR's knowledge kernel are performed in the joint fact finding phase.

As seen in Figure 13.1, MEDIATOR's design philosophy is aimed at creating an ontological design of a dispute mediation meeting environment, rather than focusing on a single module to solve portions of a problem. We hope that through the design and implementation of a system that tackles a broader range of problems, more insight into problem-solving strategies can be obtained. This, inevitably, required some simplification decisions to be made within the various modules. Each of the steps in Figure 13.1 will be discussed at some length in the following section.

System validation. To partially verify that the problem-solving mechanism embedded in the MEDIATOR framework is valid, and generally applicable to other problem domains that require managing resources among multiple parties, a validation exercise was conducted. The validation was used to verify that the prototype system was general enough to encompass other types of siting cases. A siting problem that was quite different from a MSW plant was sought, and the validation exercise was completed using the selection of factor weights for the siting of recreational playing fields in an urban area. We used the same source expert in this validation exercise, thus providing only a partial system validation. The exercise, however, was quite successful, and further details of the validation procedure are discussed at length elsewhere (Lee, 1989).

Developing the Prototype System

After the reverse knowledge engineering exercise, to develop a working prototype system with the components shown in Figure 13.1, the procedure was to: (1) make some preliminary hardware and software choices; and (2) make choices about specific analytic methodologies to be used for each system component.

Our idea was to explore the possibility of implementing our system in a PC environment using existing software packages in order to prove its portability and usability in an actual meeting environment. We hoped to show that our task, although complicated, was not going to be an impractical experiment that could exist only in a university lab. The system should be implementable with the use of straightforward methodologies and practical software packages. It

ENVIRONMENTAL DISPUTES

MEDIATOR Meeting Environment	Background Knowledge Acquisition
	Participant Identification
	Siting Factor Determination
	Preference Quantification
	Data Analysis
	Conflict Detection
	Conflict Interpretation
	Conflict Resolution

RESOLUTION

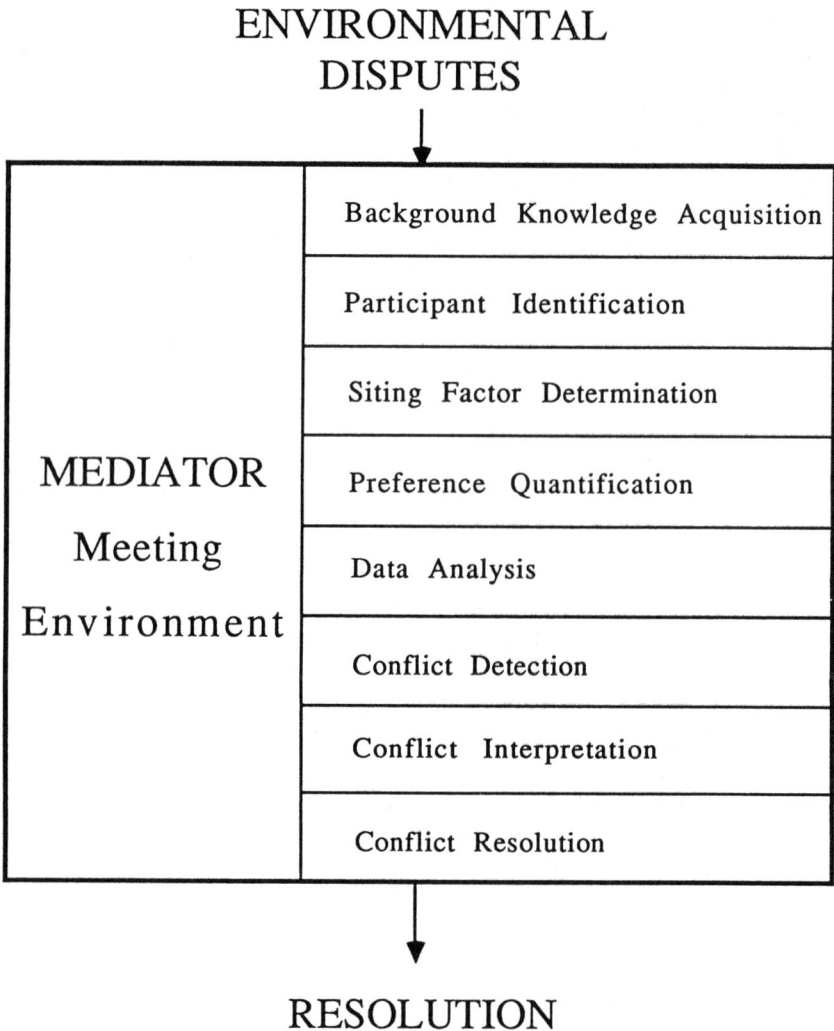

Figure 13.1. Taxonomy of mediation function for MEDIATOR meeting management environment.

was also clear that the use of an expert system shell in isolation would not be sufficient for the complex analytic tasks to be incorporated. We therefore selected to implement MEDIATOR, using a variety of linked software packages, on an Apple Macintosh II with 1 megabyte of RAM, 40 megabytes of hard disk capacity, and a math co-processor. The selected software packages included Neuron Data's Nexpert Version 1.0, Microsoft Excel Version 1.5, Systat, Cricket Graph, MacPaint and Tempo.

By making such a commitment, we inevitably were confined by some design constraints. Not surprisingly, the hardware platform and software packages left something to be desired. For example, the processing speed of the Mac-II is not fast enough, so that whenever a value is changed within a spreadsheet that needs to be propagated to other parts of the program, a lot of time is spent waiting for Excel to complete the recalculation. Due to RAM limitations, this machine doesn't support multitasking and memory paging, and therefore the number of files that can be opened at the same time is limited to only a few.

Nexpert Version 1.0 is a major contender in the object-oriented expert system shell market. It comes with a very powerful inference mechanism; yet, from our experience, it is very limited in its capability in knowledge acquisition, dynamic object inspection, explanation, and capability to interface with Excel. In particular, its lack of easy knowledge acquisition capability made creating many of the rules a tedious process. Nexpert does not allow variables to appear in dialogue windows, creating yet another problem in the user interface. Systat doesn't write its analysis outcome to syslink files, that is, files that can be retrieved and manipulated by other packages under the Mac-II operating system. This shortcoming significantly inhibited to what extent we could automate MEDIATORS's entire analysis process. Cricket Graph doesn't do automatic labeling; that is, we can get a bunch of dots nicely plotted on two axes; yet in order to find out which dot represents which data point (participant), we need to go back and manually label each dot.

Such problems seem inevitable in today's computing environment, but we believe that they will soon be phased out with the newer and more powerful software tools being released each day. It will, clearly, become easier to "take along" such computing tools to a meeting, and for professional mediators with little computing expertise to use them.

Overall, we selected the use of expert system technology to address four of the MEDIATOR functions: (1) background knowledge acquisition; (2) participant identification; (3) conflict interpretation; and (4) conflict resolution. Where more computing power was needed, we selected Excel (for siting factor determination and preference quantification). Where extensive data analysis was required, a statistical package, Systat, was used. This combining of a variety of software application packages gives MEDIATOR the flavor of a *hybrid* or *coupled* system.

The Mediator Components

Each of the system components shown in Figure 13.1 is described in more detail in the following sections. More extensive discussions of the choice of methodologies and implementation may be found in Lee (1989).

Background Knowledge Acquisition

The first module in the pre-meeting, fact-finding phase involves acquiring background information regarding the project and the candidate sites. This background information is collected using Nexpert, the selected expert system shell. Different states may have different regulations, and different types of projects may involve different sets of siting factors. Therefore, MEDIATOR needs to know where the candidate sites are and what type of project is under consideration. For the candidate facility, MEDIATOR asks for details about power plant type and size. For each candidate site, MEDIATOR solicits site specific information including its name, current zoning classification, and whether it involves the use of wetland, etc. This type of site specific information is required for the participant identification component, since, for example, a representative of the Army Corps of Engineers might be suggested if the site involves wetlands.

Participant Identification

Most environmental dispute resolution textbooks tell us that, if we look at environmental issues as a whole, four types of people, at minimum, need to present their opinions in a negotiation: regulatory agencies, developers, resident representatives, and environmental groups (Burgess and Smith, 1983; Raiffa, 1982; Sullivan, 1984; Bingham, 1986; Mernitz, 1980). Among these, the regulatory agencies and developers are relatively easy to identify, while identifying eligible environmental groups and resident representatives needs some elaboration. Yet real-world situations quickly get more complicated.

Ideally, we wish to design a mechanism that will identify just the right combination of participants; that is, all the parties that have an interest in the project and have the power to influence the decision making process are invited, and yet the total number of participants is constrained so that a negotiation to strike consensus is possible. This means that after obtaining information about the parties already present at the initial meeting, MEDIATOR must be able to detect missing parties, and also to suggest ways to prune over-represented interests.

There is simply no algorithm that can generate a correct solution to this problem. The best that we can get is a plausible expert judgment, an approximation to a solution, if any exists. This kind of problem solving task is exactly what expert systems need to explore. In the design of MEDIATOR, an expert was brought in to help construct this module. The rules used were also supplemented by information from textbooks.

In the real world, every siting exercise is a special case. The parties that would be involved in the siting of one type of facility at one specific location will definitely be different from those in other settings. Proposing a knowledge base that, in all kinds of siting exercises, would provide a complete list of all possible parties that might influence the siting process provides a classic example of combinatorial explosion (Feigenbaum, 1985). Since it is unlikely that we can build a knowledge module that can handle all types of facility siting participant identification tasks, nor would it be advisable to build MEDIATOR so that it can only tackle one special case, we need to cut the line somewhere between the two extremes. This need for compromise prompts the construction of the ideal setting.

As mentioned above, the background information acquisition module acquires project-specific and site-specific information to determine an ideal combination of participants, which we will call the ideal setting. Likewise, the participant identification module collects participant-specific information for those who are currently in the meeting room. The set of participants actually present at the meeting, which will be called the actual setting, is constantly compared with the ideal setting. MEDIATOR will make suggestions as to how to tailor the actual setting to fit the ideal setting.

The ideal setting of the participant constituency is composed of a static setting and a dynamic setting. The static setting is situation-independent. It is not affected by any special situation, and is a composition of the four main types of participants mentioned above. No matter what type of project is to be sited, in whichever state, this base setting will not change. The dynamic setting is situation-dependent. It is determined by the project-specific and site-specific information acquired in the background information acquisition module. For example, the dynamic setting for a MSW project to be sited in California may be composed of: (1) a local planning/zoning representative, if any of the candidate sites is not currently zoned for industrial use; (2) a California Energy Commission representative, if the project is sized over 50 megawatts. This approach of separating the static and dynamic settings is ideal for determining a plausible constituency of meeting participants in that the static setting is general enough to handle most siting cases, while the dynamic setting can be specific enough to take care of special situations.

After the ideal setting is constructed, the participant identification module is activated, and information about each potential participant is entered into the system. MEDIATOR then compares the ideal and actual settings, and determines which interests are over-represented and need to be pruned, as well as which interests are under-represented and need to be added. In both cases, suggestions for actions, developed from rules given by the source expert, are produced by MEDIATOR. For example, a suggestion might be to trim the number of resident representatives by qualifying petition, with groups that can solicit over 100 signatures on a petition remaining in the negotiation. The overall flowchart for this process is given in Figure 13.2

The user, of course, is free to accept or reject the advice given by MEDIATOR, and may conclude with a final composition that does not correspond to MEDIATOR's recommendation. If a perfect fit is not possible (not unusual in real-world situations), an additive failure index is assigned to the actual setting. If this index goes beyond a certain threshold, MEDIATOR will conclude that there is little point continuing a negotiation with this composition of participants, and will offer possible alternatives to negotiation.

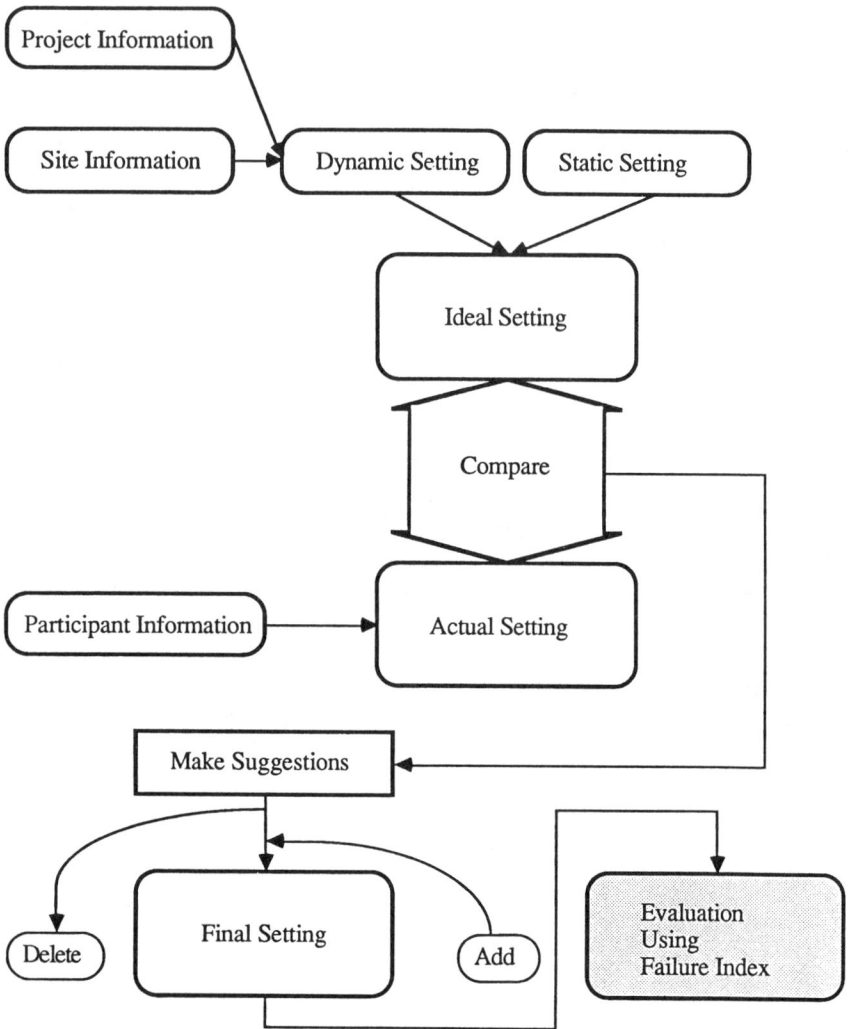

Figure 13.2. Flowchart for the background information acquisition and participant identification modules.

The knowledge representation in this module, as in the background information acquisition module, takes the form of production rules, while every entity (e.g., site, participant), is represented as an object with attributes. For example, Project is an object with attributes of State, Size and Type. Examples of production rules used in this module are:

```
IF      (Preconditions)
        Add_More_Participants is rejected, and
        New_Participant_Information_Acquired is confirmed;
THEN    (Hypothesis)
        Module_7_Completed is confirmed;
AND     (Actions)
        WRITE to a file PART each person's name, and
        WRITE to another file PART1 some other information.
```

Siting Factor Determination

After identification of the participating parties (appropriately represented or not), MEDIATOR will start to construct a list of siting factors as the next step on the meeting agenda. The system should be able to figure out an initial set of siting factors for the designated project from the background information acquisition session. The reason for giving out a set of pre-determined siting factors is that, when later adopting the decision analysis approach to quantifying the participants' preferences and positions, certain independence conditions will need to be met. MEDIATOR thus provides a set of siting factors that, according to the given background information on the sites, are likely to fulfill these independence conditions. For the case study of a MSW facility in California, this siting factor list was adapted from an actual list used by a project developer. For this specified list, MEDIATOR also provides the scaling of the factors and the estimated "best level" and "worst level" for each factor.

It is likely, however, that some participants will feel strongly that an additional factor needs to be incorporated; some may feel that certain factors must be replaced by others; some may dislike the wording of a particular factor. MEDIATOR therefore provides the meeting participants with the ability to add, delete, and replace items in the given list of siting factors.

The list of siting factors is presented to the participants in Microsoft Excel format as shown in Figure 13.3. The hard copy of this list might serve as a single negotiating text (Bacow and Wheeler, 1984) that circulates among the participants. Each person can make changes to the list and discuss the changes with other parties until an agreed-upon list is produced. The spreadsheet list is then edited according to the changes made on the hard copy.

FACTOR	CODE	DESCRIPTION	TYPE OF IMPACT	UNIT
1	X1	Parcel Size	Economic	Acres
2	X2	Access to Highway	Economic	Miles
3	X3	Proximity to Transfer Station	Economic	Miles
4	X4	Air Quality	Environmental	ppm HNO3
5	X5	Utility Service	Economic	Miles
6	X6	Zoning	Socioeconomic	Industrial Zoning
7	X7	Political Climate	Public Attitides	Months
8	X8	Construction Cost of Project	Economic	$ Millions
9	X9	Water Quality	Environmental	f t

BEST-LEVEL	WORST-LEVEL	FINE POINTS
2	50	Permanent land use
0	10	Proximity to a Highway Interchange.
1	40	Within Economic Haul to Transfer Stations.
0.02	2	NOx react with H2O in the air produces HNO3.
0	10	Proximity to Transmission Lines.
1	0	YES or NO answer whether it's industrial zoned.
2	24	TIME required to pass a regulation or ordinance.
200	1000	Construction Cost, Also considered Buildability Issue.
50	5	Distance between the incinerator and ground water table.

Figure 13.3. The list of siting factors presented by MEDIATOR.

Preference Quantification

Having determined a set of siting factors, MEDIATOR needs to obtain information about each participant's attitude toward the factors in terms of his/her initial weighting scheme. This scheme can be represented as a weight vector that takes the form:

$$W_i : \{W_{i1}, W_{i2}, W_{i3}, ..., W_{ij}\},$$

where i represents an individual that participates in the meeting, and j represents the number of siting factors considered.

The design of MEDIATOR at this point is intended to be implemented as a real-time, interactive aid for the actual mediator. The mediator, with the aid of the system, would sit down with each participant for a preference quantification session. The selected analytic tool adopted for this module is decision analysis. Decision analysis has the following advantages:

1. Decision analysis takes into consideration the decision maker's subjectivity by assessing his/her risk attitude, which is a conceptually straightforward and reasonable way to quantify subjectivity.

2. Decision analysis uses a simplified approach to assess pairwise weight assignments through direct value tradeoffs. This approach is less tedious when compared with the all pairwise weight assessment proposed by Findikaki (1986).

3. Decision analysis has an analytical framework that makes the underlying assumptions explicit.

4. Uncertainty may be addressed using probability theory.

5. The decision maker is forced to think hard about the problem (Keeney, 1980).

The common practice in decision analysis involves elaborate query sessions geared at soliciting preferences, values, risk attitudes, utilities, and consistency checks. Much research has been devoted to developing programs that automate this process, incorporating fairly elaborate procedures (Woodward-Clyde Consultants, 1979; Keeney and Sicherman, 1976; Rowley, 1989). Since we do not intend to re-invent these programs, we have chosen a less elaborate approach in programming this module, using a subset of the procedures suggested by Keeney. Microsoft Excel was selected as the implementation tool.

Each participant has one spreadsheet designated to solicit his/her weight vector. His/her name, obtained from the previous participant identification module, will appear at the beginning of the worksheet to mark such a designation. All the spreadsheets perform identical operations. In brief, each participant first selects the siting factor on the list that is the most important according to his/her opinion. All the other factors will be compared with this one (see Figure 13.4). The participant must then determine how important the other factors are, relative to the selected factor, and the utility of this value assignment is then calculated through traditional lottery procedures (see Figure 13.5). A final step of obtaining the weight of the most important factor is also carried out using a lottery (see Figure 13.6). Finally, the weights are normalized to sum to 100. The operation in the preference quantification module is summarized in the flowchart illustrated in Figure 13.7.

A valid decision analysis operation requires elaborate independence checks and consistency checks to guarantee meaningful outcomes. While implementing MEDIATOR's preference quantification module we made the assumption that these criteria are already met. As a matter of fact, computer programs developed by Rowley (1989) and Woodward-Clyde Consultants (1979) have implemented most of these checking routines. MEDIATOR can always be expanded using the designs of these existing programs if more elaborate solutions are found to be required. At this point of time, as pointed out earlier, our research concern was to address a broader range of issues rather than focusing on overly detailed implementation of each component.

```
┌────────────────────────────────────────────────┐
│ PARTICIPANT  2:                                 │
│ NAME:                                           │
│             Mr. Johnson                         │
│                                                 │
│                                                 │
│ The  MOST  IMPORTANT  siting  factor:           │
│ Enter the number of the factor from Column B    │
│ of the SitingFactor worksheet in the cell below:│
│                    4                            │
├────────────────────────────────────────────────┤
│ CODE FOR THE WEIGHT OF THIS FACTOR:             │
│                   w 4                           │
│                Air  Quality                     │
│                                                 │
│                                                 │
└────────────────────────────────────────────────┘
```

Figure 13.4. Each person will determine a most important factor on his/her designated spreadsheet.

Data Analysis

The final product from the previous module is a vector of weights for each participant. Combining the weight vectors of each and every participant, a weight matrix is constructed as a numerical description of the current conflict situation. Several multivariate analysis techniques are applied to this weight matrix. Again, not wishing to reinvent any wheels, an existing statistical package, Systat, is used by MEDIATOR.

In practice, MEDIATOR uses cluster analysis to first partition the participants into two or more clusters. Then a Classification and Regression Tree (CART) procedure is used to determine the two or three attributes that are most significant in explaining such a partition. A two dimensional graph, as shown in Figure 13.8, is the result of such an operation. Participants A to K are partitioned into two clusters due to the variation present in attribute 3 and attribute 5. Points M1 and M2 represent the mean point of each cluster. The negotiation task to be accomplished here becomes pushing cluster 1 towards M2 and cluster 2 towards M1, as closely as possible.

This visual representation is a great way to help a mediator comprehend a conflict situation. Cluster graphs like Figure 13.8 convey the sense of difference clearly and make comparisons straightforward. For both the participants and the mediator, these graphs highlight important differences and points of conflict so that the discussion may focus on central issues and effective negotiation.

DETERMINE THE VALUE OF u4 : Air Quality

```
                                            _____  1      u4 (  0.02  )
                                           |   0.5
                                           |
              u4 (   1.2   )   0.5 _____|
                                           |
                                           |
                                           |_____  0      u4 (   2   )
                                                0.5

                                            _____  0.5    u4 (  1.2  )
                                           |   0.5
                                           |
              u4 (   1.7   )   0.25 _____|
                                           |
                                           |
                                           |_____  0      u4 (   2   )
                                                0.5

                                            _____  1      u4 (  0.02  )
                                           |   0.5
                                           |
              u4 (   0.75   )   0.75 ____|
                                           |
                                           |
                                           |_____  0.5    u4 (  1.2  )
                                                0.5
```

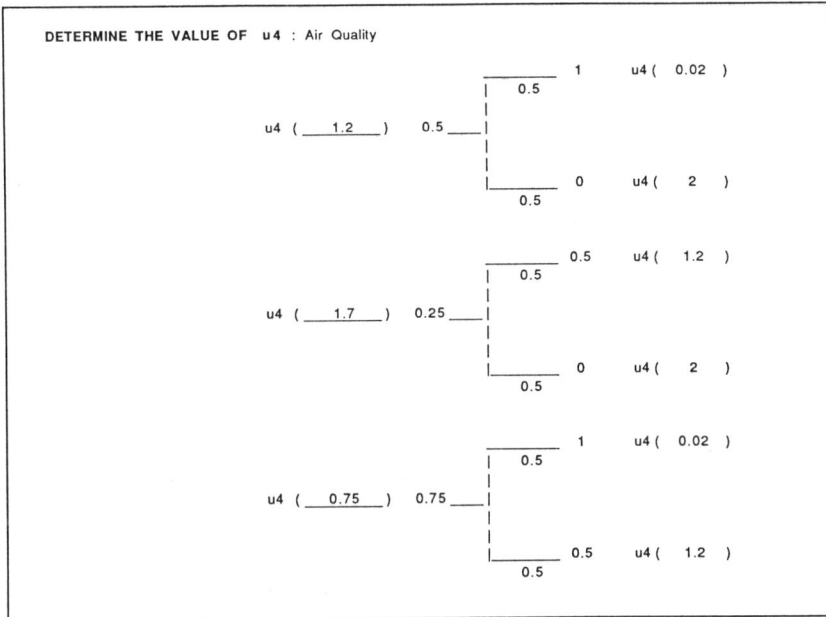

Figure 13.5. The three lotteries used to solicit Mr. Johnson's utility curve.

In practice, the Systat DATA module is used to prepare the weight matrix into a Systat readable format; then the Systat CLUSTER module is applied to this formatted file to perform cluster analysis. The objective of this stage is to detect a cluster pattern among all the participants' preferences as reflected in the weight matrix. In the initial research, the number of clusters was restricted to 2. A typical result may be seen in Figure 13.9, which illustrates the dendogram obtained after applying the average linkage agglomerative hierarchical cluster algorithm to our weight matrix. The selection of a particular cluster algorithm is intrinsically a matter of taste and subjective judgment. Many analysts favor the use of the average linkage method because of its ability to recover the underlying cluster structure based on internal criterion measures, and it was adopted for MEDIATOR. The cluster analysis information is then sent over to CART for a second stage analysis.

CART analysis works similarly to discriminant analysis (Breiman, Friedman, Olshen, and Stone, 1984). The methodology is used to determine important factors that result in a given partition within a training group; so that later on these factors can be utilized to perform a classification on any similar group that is being studied. CART treats the classification operation as branching on a tree. This binary tree classifier is a simple, yet powerful, method for classification. For complex data with many independent variables, the binary trees produced by CART can have error rates that are significantly lower than

```
DETERMINE THE VALUE OF  w 4  :
                                    _____  All  attributes
                                  |        P       at BEST level.
                                  |
    X4                            |
    at BEST level,    _____   |
    All others at WORST           |
    level.                        |
                                  |
                                  |
                                  |_____  All  attributes
                                       1 - P      at WORST level.


                   P    =    0.75  ---->   w 4    =   0.75
```

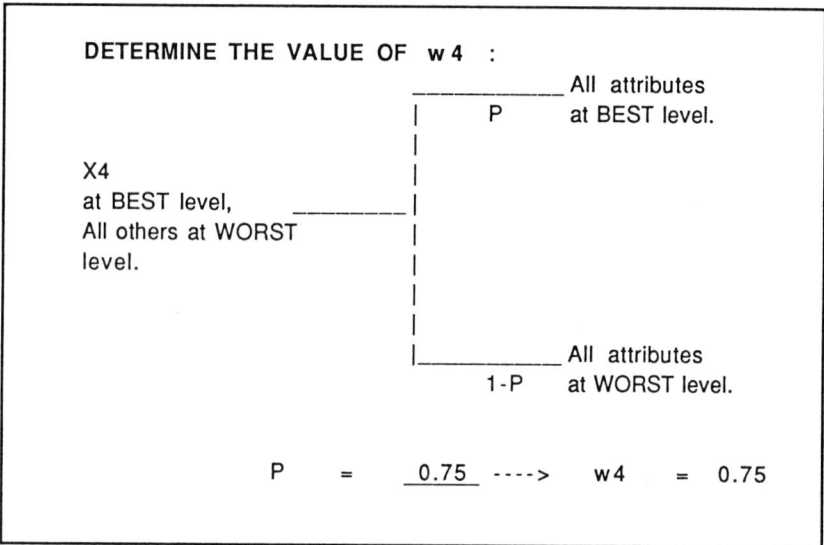

Figure 13.6. Lottery to solicit the value of factor 3's weight $W_{Johnson,3}$.

those produced by the usual parametric procedures (Komor and Wiggins, 1989). The CART analysis performed by MEDIATOR is a subset of the full-blown CART, used to determine the two or three most important factors that result in the two- or three-cluster partition detected in the cluster analysis. It is implemented using Microsoft Excel. The operations within the data analysis module are summarized in Figure 13.10. More details about the data analysis component may be found in Lee (1989).

Conflict Detection

The most important reason for adopting a visual representation like that shown in Figure 13.8 lies in its ability to assist the mediator in conflict detection. Remember that this graph is generated by applying several statistical techniques. If two groups of participants show significant preference variation from each other due to variations in two attributes, this might be a point to start a mediation attempt. Specifically, the layout of the clusters gives abundant information for conflict detection and interpretation.

MEDIATOR assists in the detection of conflicts by informing the user of the 2 participants who are most diverse in their opinions from the mean point of each of the clusters. For example, the person in cluster 1 that is least in agreement with the people in cluster 2 is identified. From each participant's preference quantification worksheet is also obtained information about their propensity to move. That is, a measurement of how willing this person is likely to be

NEXPERT

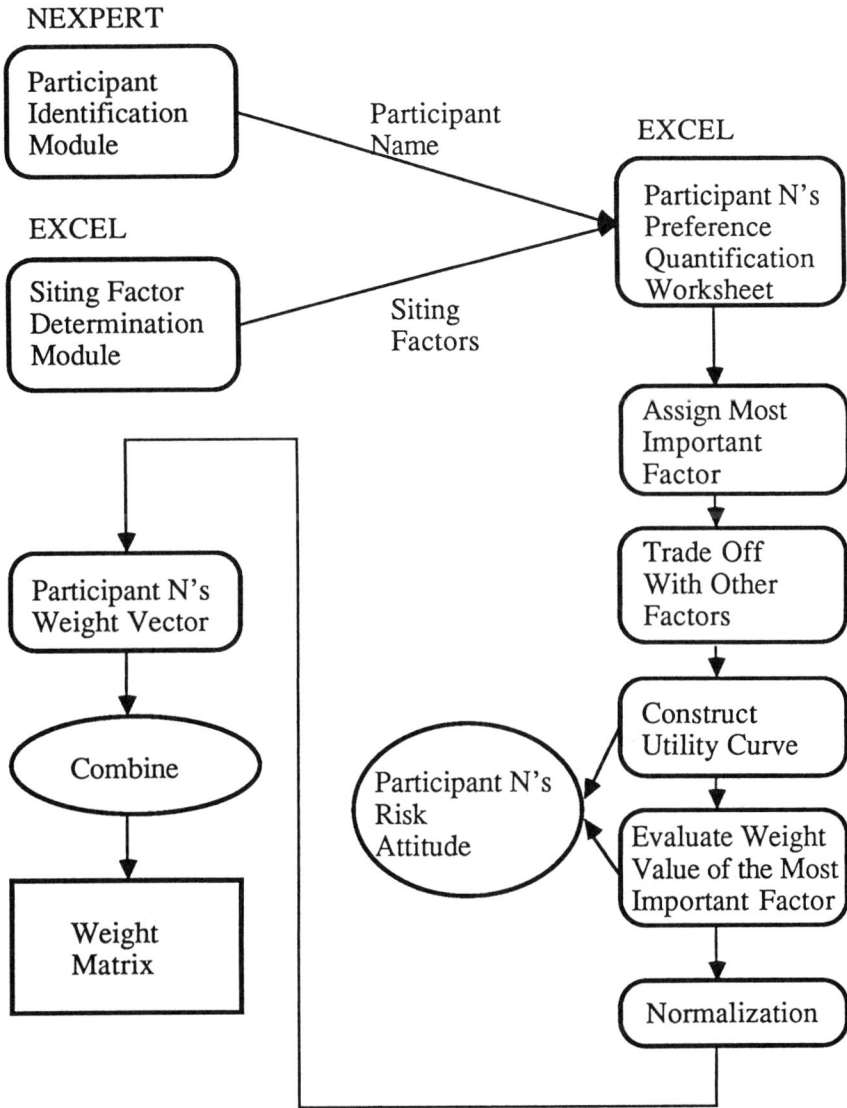

Figure 13.7. Flowchart for MEDIATOR'S preference quantification module.

to move from an original assignment to a new one. This exercise is helpful to the practicing mediator in that, while managing a meeting, it is important to know how strongly each person holds to his/her position. Since the data analysis developed here concentrates on the most important distinguishing factors in disagreement, measuring the propensity to move between these factors

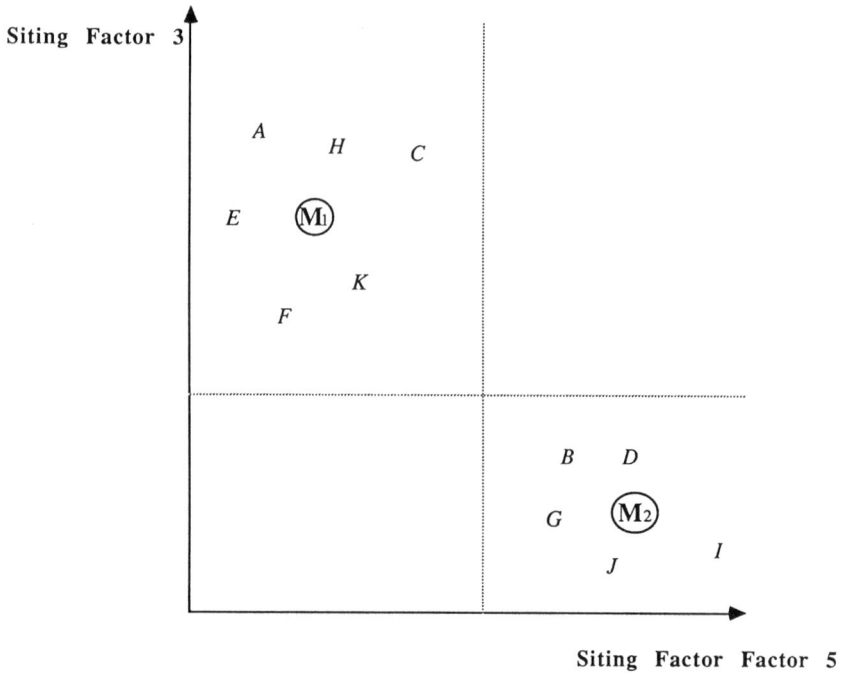

Figure 13.8. Partitioning participants into clusters according to two attributes.

provides an indication about the degree to which each person might be willing to reconcile. For those who hold extreme positions with little propensity to move, the mediator might wish to seek clarification of positions.

Conflict Interpretation and Conflict Resolution

There are various ways to deal with the detected conflicts. Much research has been devoted to devising analytical approaches that would come up with a "best" allocation of resources that would "maximize total gains" for all the participants. Most of the research appears to be centered around multiagent planning and analytical arbitration issues (Raiffa, 1982; Muscettola and Smith, 1987; Zahedi, 1986; Kirkwood and Sarin, 1972; Balson, 1982; Bodily, 1980). The problem with these approaches is that they are oriented towards an arbitrated outcome, and a negotiation approach for conflict-resolution is not addressed.

In this research, our approach is to associate yet another knowledge base with MEDIATOR for conflict interpretation and conflict resolution. We term this module the kernel knowledge base because it is the core module of the

DISTANCE METRIC IS EUCLIDEAN DISTANCE
AVERAGE LINKAGE METHOD

TREE DIAGRAM

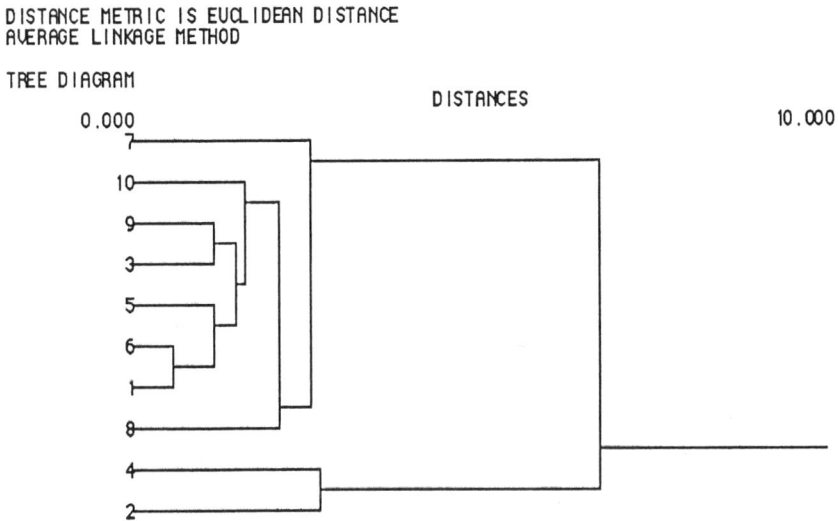

Figure 13.9. Dendrogram illustrating the current clustering partition.

entire MEDIATOR system. Conflicts detected by the meeting control protocol are analyzed and interpreted according to the information previously obtained by the preference quantification module. Then a set of production rules provides strategies and plans to tackle these conflicts.

MEDIATOR's conflict interpretation module searches through the database that stores all the value trade-off information, retrieves the items pertaining to a detected conflict, then sends them over to the rule-based conflict resolution module which generates suggestions for mediating the conflict situation. The source expert, with additional rules from the literature, provided the main body of knowledge for the conflict resolution module. An example of a suggestion from MEDIATOR might be: *"Ask the two participants with the most diverse opinions to trade-off between their assignments on factor 4. See if there is another factor that displays similar discrepancies that can be used by these two people to bargain about each other's factor 4 assignment."* Another might be: *"Start working with people who show low propensity to move values, i.e., people that are likely to make changes on their factor 1 and factor 4 assignments. In this case, they are participants 1, 6 and 10. If the mediator successfully encourages these people to change their assignment, this might become a role model for other participants."*

Negotiation under the auspices of MEDIATOR is a continuous, often iterative process. The system makes suggestions to the real-life mediator as to what strategies would be appropriate. The mediator, as well as the meeting participants, can always reject a suggestion. Then MEDIATOR will look at other

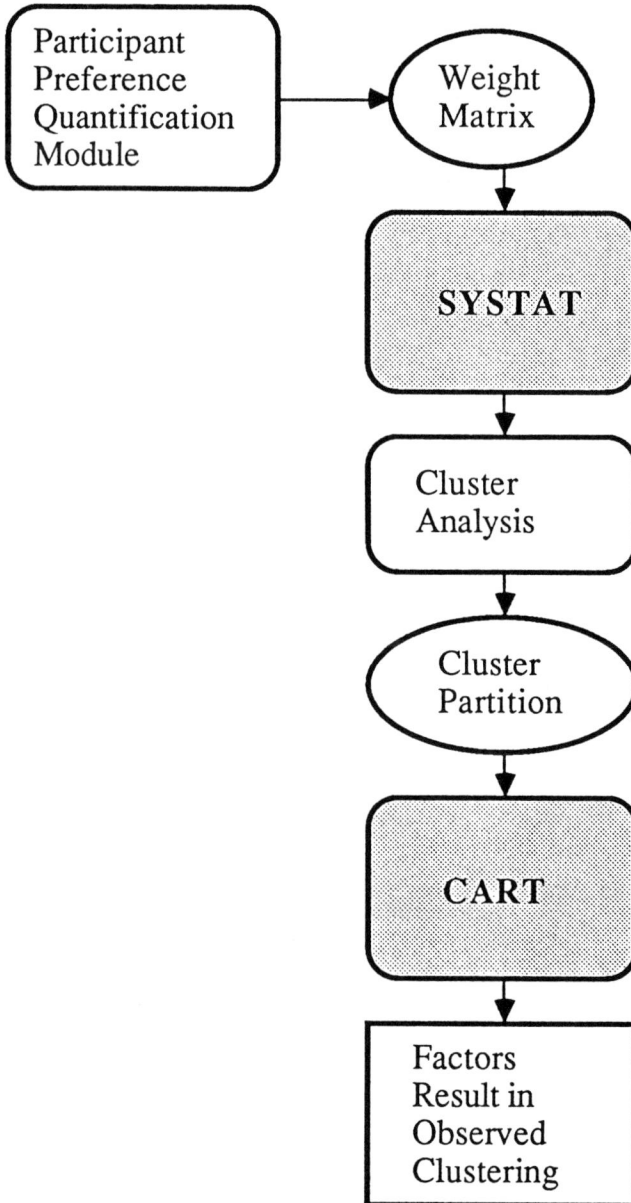

Figure 13.10. The flowchart of MEDIATOR's data analysis module.

value trade-offs to provide another suggestion. This process, as illustrated in Figure 13.11, will iterate until an agreement (tolerable under certain statistical criteria) is reached or a deadlock is met.

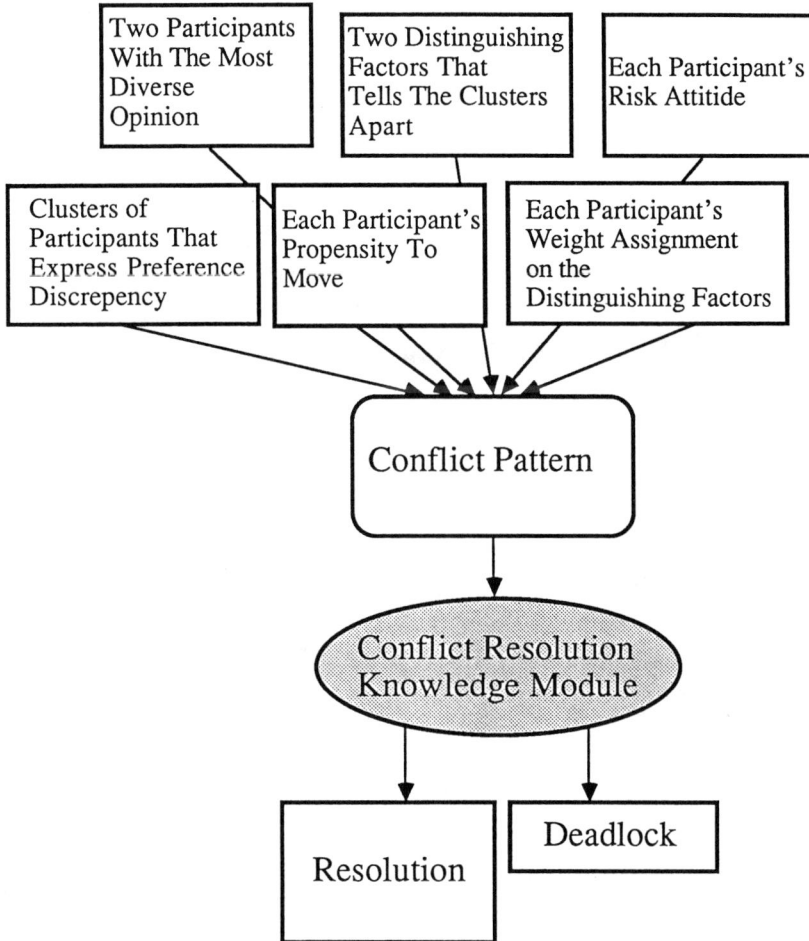

Figure 13.11. The flowchart for MEDIATOR's conflict resolution module.

Termination Conditions

Suppose that, after the mediator's effort, some participants have agreed to change their weight assessments on specific factors. They can go back to their individual worksheets and make changes to the section where they specify the value trade-offs among factors. Every change a person makes will result in a different weight vector, and thus a different ranking on site preferences. When the system recognizes a point where every participant has selected the same site as his/her first choice, the negotiation is declared completed. In other situations, the conflict will not be mediable, and deadlock will be the conclusion. An easy example of likely deadlock is when a participant is holding an extreme position without compromise. MEDIATOR's conclusion in this case will appear as in Figure 13.12.

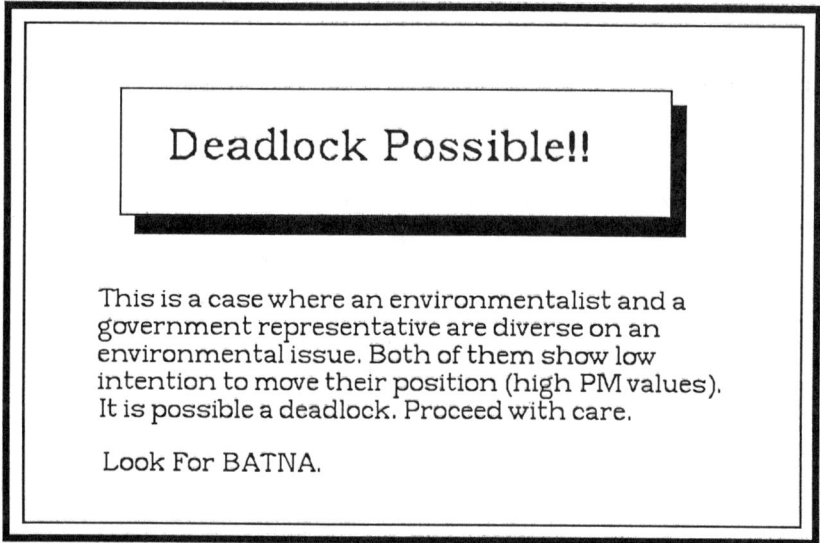

Deadlock Possible!!

This is a case where an environmentalist and a government representative are diverse on an environmental issue. Both of them show low intention to move their position (high PM values). It is possible a deadlock. Proceed with care.

Look For BATNA.

Figure 13.12. MEDIATOR detected deadlock possibility from the conflict situation.

Directions for Further Research

MEDIATOR is not designed to solve all environmental dispute resolution problems, but it is a successful prototype "hybrid" expert system. All of the components shown in Figure 13.1 have been implemented, and the system has been partially validated by using a very different case example. However, much work remains to be done. Future MEDIATOR research projects might be focused in the following directions:

1. Enhance the knowledge base in the participant identification module to more fully account for the siting of a larger variety of facilities at different locations.

2. Keep adding rules to enhance the kernel knowledge base so that its conflict identification and conflict resolution tasks cover more ground.

3. Write more code to automate more completely the linking between programs and files, so that the user can worry less about navigating between the different packages.

4. In addition to ranking and choosing among candidate sites, take into consideration SITE-NO SITE alternatives (e.g., consider a *no-build* alternative where people might express sentiments like, *"we just don't want any nuclear power plants"*).

5. Make use of case-based reasoning methodologies to refine the system's problem solving behavior.

6. Consider computer-based multi-user communication, like the COLAB project conducted by XEROX-PARC, where each user is equipped with a WYSIWYG (What You See Is What You Get) terminal. Any individual's thoughts and contributions to a discussion topic appear simultaneously on the other individuals' screens. This may be the next generation of teleconferencing, and MEDIATOR could benefit from ideas in this area.

7. Port a portion of the system that is already implemented over to other software and hardware platforms.

To make sense out of such an unstructured, fuzzy problem domain as a mediation problem, special steps needed to be taken to transform the problem into a more structured form that allowed for the development of a decision support system. In MEDIATOR, a mix of methodologies including decision analysis, multivariate statistical analysis, and CART were used to make the problem more clearly structured. The expert system approach was then also adopted into the "hybrid" system because of its capability to provide plausible, rather than optimal, solutions. The expert system approach proved to be adequate for this type of problem domain precisely because of the domain problem's lack of precision and correctness of solution.

220   Yi-Chin Lee and Lyna L. Wiggins

References

Bacow, L. and M. Wheeler, 1984. *Environmental Dispute Resolution,* Plenum Press: New York.

Balson, W.E., 1982. "The Foundations of the Design of Agreement: A Quantitative Theory of Arbitration," Ph.D. dissertation, Stanford University, Stanford, California.

Bercovitch, J., 1984. *Social Conflicts and Third Parties: Strategies of Conflict Resolution,* Westview: Boulder, Colorado.

Bingham, G., 1986. *Resolving Environmental Disputes, A Decade of Experience,* The Conservation Foundation: Washington, D.C.

Bodily, S.E., 1980. "A Delegation Process for Combining Individual Utility Functions," *Management Science,* 25(10):1035-1041.

Breiman, L., J. Friedman, R. Olshen and C. Stone, 1984. *Classification and Regression Trees,* Wadsworth: Belmont, California.

Buckle, L.G. and S.R. Thomas-Buckle, 1986. "Placing Environmental Mediation in Context: Lessons from Failed Mediations," *Environmental Impact Assessment Review,* 6(1):55-70.

Burgess, H. and D. Smith, 1983. *The Use of Mediation: The Bryton Point Coal Conversion Case,* Schenkman Books, Inc.: Cambridge, Massachusetts.

Cormick, G.W., 1973. "Environmental Mediation: An Action Proposal," *Technical Report,* Community Crisis Intervention Center, Washington University, St. Louis, Missouri.

Cormick, G.W. and J. McCarthy, 1974. "Environmental Mediation: A First Dispute," *Technical Report,* Community Crisis Intervention Center, Washington University, St. Louis, Missouri.

ERCDC (California Energy Resource Conservation and Development Commission), 1978. *The Siting Manual,* EDAW Inc.

Feigenbaum, E., 1985. Course notes, Introduction to Artificial Intelligence, (CS123), Computer Science Department, Stanford University, Stanford, California.

Findikaki, I.T., 1986. "An Expert System for Site Selection," *Proceedings of the ASCE Conference on Expert Systems in Civil Engineering,* ASCE.

Fisher, R. and W. Ury, 1981. *Getting to Yes,* Penguin Books: Middlesex, England.

Folberg, J. and A. Taylor, 1984. *Mediation: A Comprehensive Guide to Resolving Conflicts without Litigation,* Jossy-Bass Publishing Inc.: San Francisco, California.

Harashina, S., 1988. "Environmental Dispute Resolution in Road Construction Projects in Japan," *Environmental Impact Assessment Review,* 8:29-41.

Hopkins, L.D., 1977. "Methods for Generating Land Suitability Maps: A Comparative Evaluation," *Journal of the American Planning Association,* 43, (October): 386-399.

Keeney, R.L., 1980. *Siting Energy Facilities,* Academic Press: New York, New York.

Keeney, R.L. and C.W. Kirkwood, 1975. "Group Decision Making Using Cardinal Social Welfare Functions," *Management Sciences,* 22(14): 430-437.

Keeney, R.L. and A. Sicherman, 1976. "An Interactive Computer Program for Assessing and Analysing Preferences Concerning Multiple Objectives," *Behavioral Science,* 21:173-182.

Kirkwood, C.W. and R.K. Sarin, 1985. "Ranking with Partial Information: A Method and an Application," *Operations Research,* 33(1): 38-48.

Kolb, D., 1983. *The Mediators,* MIT Press: Cambridge, Massachusetts.

Komor, P.S. and L.L. Wiggins, 1989. "Discrete Choice Methods: An Introduction to Tree-Structured Classification," *Regional Science Review,* forthcoming.

Lake, L., 1980. *Environmental Mediation: The Search for Consensus,* Westview Press: Boulder, Colorado.

Lee, Y., 1989. "MEDIATOR: An Expert System Approach to Facilitate Environmental Dispute Resolution," a dissertation submitted to the Department of Civil Engineering, Stanford University.

Mernitz, S., 1980. *Mediation of Environmental Disputes - A Source Book,* Praeger Publishers: New York, New York.

Moore, C.W., 1986. *The Mediation Process,* Jossy-Bass Publishers: San Francisco, California.

Muscettola, N. and S.F. Smith, 1987. "A Probabilistic Framework for Resource- Constrained Multi-Agent Planning," *Proceedings for IJCAI 87, International Joint Conference on Artificial Intelligence,* IJCAI, August.

O'Connor, D., 1978. "Environmental Mediation: The State-of-the-Art," *Environmental Impact Assessment Review,* 2:9-17.

Raiffa, H., 1982. *The Art and Science of Negotiation,* Harvard University Press: Cambridge, Massachusetts.

Rowley, J., 1989. "Using the Beta Distribution in the Analysis of Risky Choices: Three Experimental Implementations of Multiattribute Utility Theory," Ph.D. dissertation, Department of Operations Research, Stanford University, Stanford, California.

Schachter, R.D. and D.E. Heckerman, 1987. "Thinking Backward for Knowledge Acquisition," *AI Magazine,* 8(3):55-61.

Sowa, J., 1988. Lecture Notes, Conceptual Structures, (CS309a), Computer Science Department, Stanford University, Stanford, California.

Sullivan, T.J., 1984. *Resolving Development Disputes Through Negotiations,* Plenum Press: New York, New York.

Susskind, L., L. Bacow and M. Wheeler, editors, 1983. *Resolving Environmental Regulatory Disputes,* Schenkman Books, Inc.: Cambridge, Massachusetts.

Susskind, L., 1987. "Negotiating Better Development Agreements," *Negotiation Journal,* 3(1):11-15.

Waterman, D.A., 1986. *A Guide to Expert Systems,* Addison-Wesley Publishing Company.

Woodward-Clyde Consultants, 1979. "Environmental Assessment Methodology: Solar Power Plant Applications, Decision Analysis Computer Program," *Technical Report ER-1070,* Electric Power Research Institute.

Zahedi, F., 1986. "Group Consensus Function Estimation When Preferences Are Uncertain," *Operations Research,* 34(6): 883-894.

PART FIVE

Expert Systems in Urban Planning: Future Research Directions

Introductory remarks by Tschangho John Kim

In the face of uncertainty and dynamic changes in urban society, the rational decision-making model that has dominated the urban planning field for decades may not be adequate in responding to continuously challenging urban problems. A typical rational decision-making model is designed to produce optimal solutions and, in most cases, such solutions neither exist nor are implementable. At the same time, artificial intelligence (AI), a newly emerging technology, can not provide all the solutions to urban planning problems by itself, although the heuristic method of expert systems (ES) attempts to produce acceptable solutions most of the time. Indeed, the rational decision-making approaches may best be supplemented by AI technology to provide more realistic and intelligent solutions to urban problems.

Unlike other chapters in this book where actual expert systems (even though most of them are prototypes) are introduced, the chapters in Part Five are intended to shed light on issues of how expert systems can be incorporated and integrated into the existing planning process in order to provide more intelligent planning solutions in the future.

In Chapter 14, Heikkila, Moore and T.J. Kim propose a method by which ES can be integrated into geographic information system (GIS) and into transportation, land use and population modeling for the provision of infrastructure. The authors begin by briefly summarizing the state-of-the-art of land use-transportation modeling by the Southern California Association of Governments in a typical metropolitan area in the USA in terms of planning. Next, they describe how a GIS model can be used to produce plans for infrastructure-based services, as well as how ES and GIS can be coupled to generate an expert geographic information system (EGIS). Finally, they proposed a new approach of incorporating EGIS into a land use transportation modeling environment. Although the chapter does not provide the exact nature of such an integrated model, it provides valuable future directions as to how expert systems may be used in conjunction with existing planning processes.

In Chapter 15, Han and T.J. Kim review past and current developments in urban information systems and provide a research agenda for the future. They begin with definitions and a topology of urban information systems. Urban information systems (UIS) are reviewed by dividing them into four distinct groups: (1) database management systems (DBMS); (2) geographic information systems (GIS); (3) decision support systems (DSS); and expert systems (ES). Next, they provide new directions for future research in three areas: (1) coupling ES and DBMS; (2) coupling ES and GIS; and (3) coupling ES and DSS. The chapter ends with a discussion of a few important future research topics in developing expert systems in urban planning that include: (1) reasoning with uncertainty; (2) knowledge acquisition and learning; (3) conflict resolution; and (4) evaluation and validation of the system.

CHAPTER 14

Future Directions for EGIS: Applications to Land Use and Transportation Planning

Eric J. Heikkila, James E. Moore
and Tschangho John Kim

A principal responsibility of urban governments is to efficiently provide infrastructure-based services to residents. The level of service derived from infrastructure depends as much on the prevailing land use and demographic characteristics of the local area as it does on the attributes of the infrastructure itself. For this reason, urban planning for the provision of infrastructure-based services is made more difficult in large urban areas that are subject to rapid change in size or character. Recognition of this basic fact is the underlying motivation for this chapter.

Unfortunately, most models of the interaction between infrastructure systems and the evolution of urban land uses and activities are quite limited. These limitations stem from a failure to address nonlinearities in production technologies, and a failure to capture the spatial relationships inherent in many service delivery systems. It is essential that such effects be accounted for if local governments are to undertake meaningful analysis of the infrastructure-based service impacts imposed by urban land use changes. This is particularly pertinent for urban areas that are undergoing a spatial restructuring of activities and land uses, because in these instances the incompatibility between current uses and inherited infrastructure systems can become most apparent.

The purpose of this chapter is to evaluate the general feasibility of incorporating expert geographic information systems (EGIS) into local government infrastructure planning models. To guard against unrealistic speculation on our part, the subsequent analysis is rooted in a particular local government modeling environment, that of the Southern California Association of Governments (SCAG). The next section describes the salient features of SCAG's current transportation modeling environment. Following this is an examination of how expert geographic information systems (EGIS) can be used to evaluate the impacts of urban change on the Southern California transportation system. The penultimate section outlines how EGIS could be incorporated directly into SCAG's modeling environment. The concluding section assesses the general feasibility of this approach.

Land Use Transportation Modeling at SCAG

As a Council of Governments, the Southern California Association of Governments is a regional planning organization responsible for developing forecasts of population and land use changes for the five counties of the Los Angeles metropolitan area. At present, SCAG's modeling efforts are supported by a conventional urban transportation modeling system, as well as a small area forecast model and a resident geographic information system. Viewed in aggregate these model components provide SCAG's staff with the capability to produce baseline population and land use forecasts that reflect the current state of the art in iterative land use and transportation forecasting applications.

SCAG makes its transportation modeling projections by combining a sequential application of its urban transportation planning (UTP) procedures with an informal set of expert assumptions concerning future land use and growth controls in the Los Angeles basin. The spatial configuration of land uses and activities that initiate the UTP system can be empirical, or can be the outputs of a small area forecasting model derived from DRAM-EMPAL (Putman, 1977). This model assigns aggregate growth estimates for the region on a top-down basis.

For any given land use configuration, a fully iterated UTP system generates an internally consistent set of predictions for trip generation, trip distribution, mode split, network assignment and level of service. These outputs from the UTP component serve as inputs to DRAM-EMPAL. Ideally this system converges to a general equilibrium forecast in which the UTP and DRAM-EMPAL forecasts are mutually consistent. This idealized state-of-the-art approach is summarized in Figure 14.1.

SCAG's conventional UTP procedures are enhanced by the Association's GIS capability. The GIS is currently used in this context as a means of storing and retrieving spatially oriented data pertaining to land use activities and the transportation network. However, neither the UTP models nor the small area forecast model nor the GIS software is sufficient to adequately model the anticipated interactions between the transportation and land use activity systems.

In particular, expert assumptions regarding the pattern of future land use controls and infrastructure investments underlying these projections are at present introduced on an informal or ad hoc basis. Consequently, the market-based interactions between the evolving urban activity system and the level of service provided by the transportation network are accounted for in SCAG's

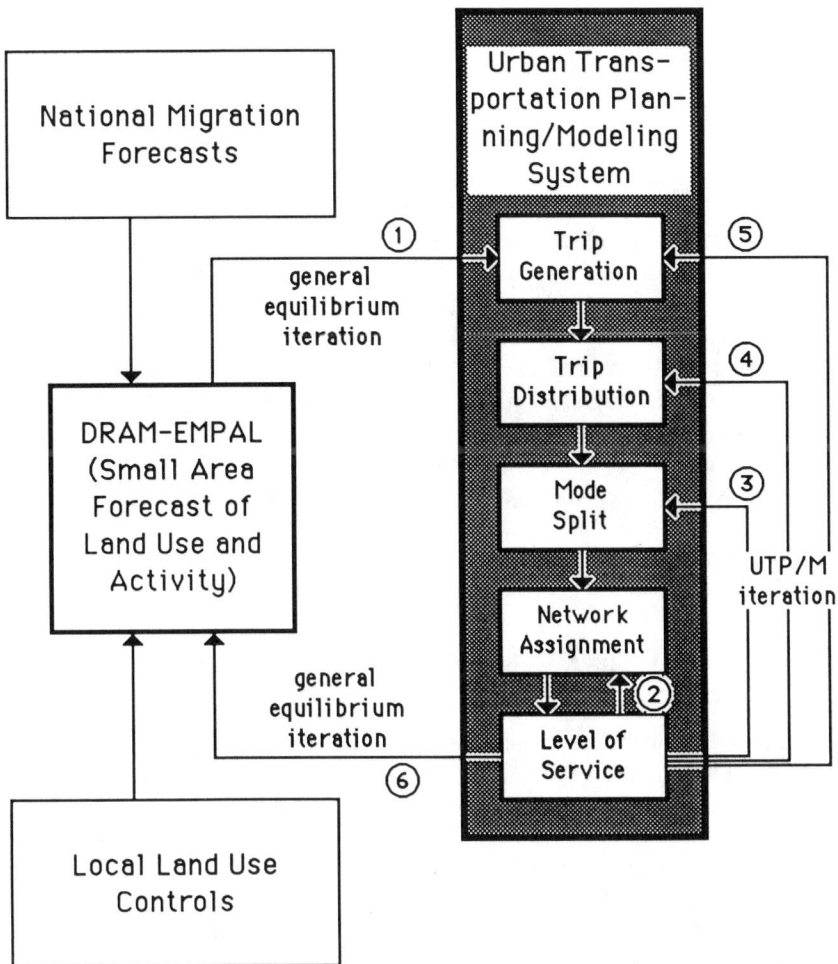

Figure 14.1. An iterative approach to general equilibrium forecasts: land use and transportation.

Draft Baseline Projection for the year 2010, but in an indirect, unsystematic way. As we shall see, an EGIS is best suited to assist with this aspect of the overall modeling framework.

Modeling the Impact of Growth on Transportation Infrastructure Using an Expert Geographic Information System

Defining Impacts

The impact of growth on infrastructure based transportation services has two principal aspects. One is the change in service levels brought about by imposing changing land uses on a fixed infrastructure system. For example, as residential densities and the distribution of employment centers changes, congestion levels will typically rise or fall by varying amounts over different network links. An aggregate index of the level of service derived from the transportation system might include median trip duration or measures of congestion throughout the system.

The second aspect of the impacts of growth on transportation based services pertains to mitigation costs. These are the costs of additional public investments required to offset what would otherwise be an unacceptable decline in transportation level of service. Most notably this would include the costs of system expansion, i.e., the construction of new transportation arteries or an increase in the capacities of existing links. If land uses are intensifying, both increased investment and congestion costs will be experienced, as congestion is traded off against additional infrastructure improvements.

Using GIS to Model the Production of Infrastructure-Based Services

The manner in which these impacts are manifested is both geographic and complex. For this reason GIS can be a useful tool for modeling these phenomena. A GIS is a unique database management system (Han and Kim, 1989; Raster, 1978) that facilitates the storage and retrieval of information on the spatial attributes of objects, and on the spatial relationships between objects. The ability of a GIS to efficiently infer these spatial relationships (such as "near" or "downhill from") is determined by the manner in which the spatial attributes of objects are represented in the database (Dueker, 1987). This inference capability distinguishes GIS from more conventional computer-aided mapping (CAM) programs (Crosswell, 1986). CAM systems merely overplot separate layers of data types and are unable to relate data across layers.

GIS is a useful tool for modeling the spatial aspects of production processes for infrastructure based urban services. Together with adjunct engineering, economic or behavioral models it can provide a very direct representation of traffic flows along transportation corridors, water flows through a network of pipes, messages across telephone trunks, or potentials along power lines. And, it can relate these representations to surrounding land uses and other neighborhood characteristics. Level of service indexes such as congestion, median trip time, water pressure, or network access, may be directly inferred from the state of the system at any time. Moreover, changes in these service levels brought about by shifts in the land use activity system may be inferred by incorporating appropriate engineering, economic or behavioral assumptions. In short, it is this capacity for complex spatial analysis that identifies a GIS production model as the spatial analogue to more standard economic models of production in an aspatial context (Heikkila, 1989).

A GIS production model comprises not only the GIS but also the adjunct models that facilitate the simulation of production flows or outputs within a specific land use and infrastructure context as defined by the GIS. Interactions between the GIS and its adjunct production models are summarized in Figure 14.2. In this diagram the endogenous level of services derived from an infrastructure system is based on the level of past investment $ in the system, as well as on the spectrum of land use activities x throughout the urban system. This relationship may be summarized symbolically by the production relation

$$s=f(x,\underline{\$}) \tag{1}$$

where an underlined variable indicates that it is exogenously determined. Similarly, the adjunct production models may produce endogenous estimates of repercusive activity shifts resulting from changing service levels. For example, as congestion increases along certain transportation links, land use activities may relocate to less congested areas. This response may be summarized symbolically as

$$x=g(s,\underline{\$}) \tag{2}$$

A partial equilibrium forecast would yield mutually consistent estimates of land use activities and transportation service levels. This partial equilibrium condition may be expressed symbolically in two ways:

$$s=f[g(s,\underline{\$}),\underline{\$}] \tag{3a}$$

$$x=g[f(x,\underline{\$}),\underline{\$}] \tag{3b}$$

Ideally, this is the type of forecast that would be generated by the SCAG modeling environment depicted in Figure 14.1. The UTP models estimate levels of service for a given land use and infrastructure configuration, and the small area forecast model addresses the spatial allocation of land use activities. However, the forecasts generated by SCAG are not yet fully iterated ones, and so cannot be viewed as partial equilibrium forecasts in the sense of Equation 3a, b.

Because the infrastructure system is exogenously determined, the GIS production model described above generates what is at best a partial equilibrium forecast. It is not sufficient for calculating estimates of efficient mitigation costs. Specific knowledge and expertise are required to determine which mitigation options are appropriate in any given context. The GIS production model in Figure 14.2 does include adjunct models that simulate traffic flows for a given

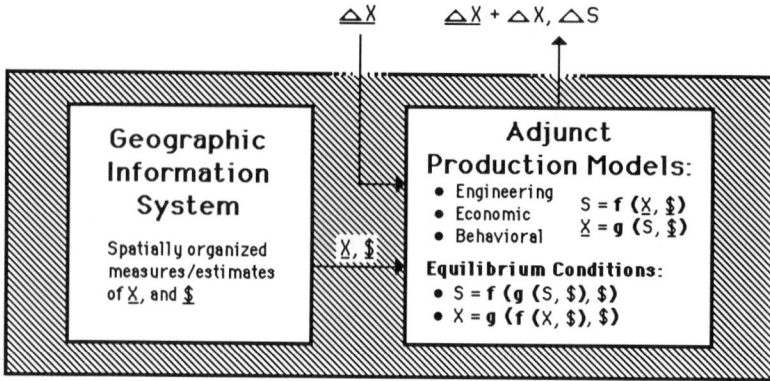

X = existing land use/demographic context

$X = g\,(S, \$)$ = endogenous land use and population consistent with S and $

$\triangle X$ = exogenous change in land use/demographic context

$\triangle X$ = endogenous activity shift

S = existing infrastructure level of service

$S = f\,(X, \$)$ = endogenous level of service consistent with X and $

$\triangle S$ = endogenous change in infrastructure level of service

$\$$ = existing infrastructure system

Figure 14.2.   A Geographic information system linked to adjunct production models: A GIS production model.

transportation network, but these model components are only used to determine the service impacts and repercusive activity shifts resulting from exogenous land use changes. They do not indicate what the most appropriate mitigation strategy might be. It is our contention that an expert system is the appropriate vehicle for incorporating such judgemental capabilities in a predictive modeling context.

General Equilibrium Forecasts: Using ES to Estimate Infrastructure Improvements Endogenously

Conveniently, the cost of infrastructure investment can also be explored in this same general framework. Ideally, we want to determine the least cost of reconfiguring the infrastructure system so that specified service levels may be upheld despite projected changes in land use activities. In order to obtain such estimates we must first determine what the appropriate infrastructure improvements are in response to declining service levels brought about by these evolving land use activities. That is, the infrastructure configuration must itself be determined endogenously. Expressed symbolically, we want

$$\$=h(x,s) \tag{4}$$

And the full set of equilibrium conditions becomes

$$s=f[g(s,\$),h(x,s)] \tag{5a}$$

$$x=g[f(x,\$),h(x,s)] \tag{5b}$$

$$\$=h[g(s,\$),f(x,\$)] \tag{5c}$$

It is easy to assert that a more general equilibrium forecast is desirable. However, in complex infrastructure systems, there are likely to be innumerable infrastructure improvements that could restore service levels. Consider, for example, all possible improvements that might be made to the transportation network in a metropolitan area. Formal and informal reasoning processes based on expert knowledge and judgement may be needed to determine reasonable bounds for the technically feasible set of alternatives. To specify a complete spatial model of impacts for spatially based services, it is therefore necessary to emulate this expert knowledge and judgement. What this requires is an efficient (heuristic) algorithm for reducing the complete set of possibilities to a few candidate choices. The optimal choice may not be contained in the candidate set. All that is known is that the optimal solution has not necessarily been excluded from this set, and that all candidate choices are likely to be good alternatives.

This is precisely the type of task that an expert system is designed to undertake (Goodall, 1985). An expert system is software designed to replicate expert human reasoning processes within the context of specific problem-solving domains. The essence of an ES is that it systematically incorporates the (useful) judgment, experience, rules-of-thumb, and intuition of human experts into problem solving. According to Waterman (1986), "an algorithmic method of conventional programming is designed to produce an optimal solution whereas the heuristic method of expert systems produce an acceptable solution most of the time." Thus the problems suitable for expert system development must accept heuristic solutions. Urban infrastructure decisions clearly fall into this category. In the context of infrastructure, identifying the optimal member of a feasible decision set is not usually a tractable exercise. Success is constrained by limited information concerning both alternative courses of action and their outcomes (Moore, 1986).

An analogy to chess-playing is helpful. For any given state defined by a configuration of playing pieces, there may be innumerable feasible moves. Yet chess masters quickly disregard all but a handful of these and explore only a few options rather intensively. What permits them to do so is a well-structured knowledge base regarding types of positions, the rules of the game, and related implementation strategies directed towards a specific objective. This (heuristic) approach is equally useful in the context of urban infrastructure and land use planning problems.

Components of an Expert System

As Figure 14.3 illustrates, there are three primary components in an expert system; a user interface, a knowledge base, and an inference engine. The user interface facilitates user interaction with the system. It translates queries from the user into specific goals for the expert system's inference engine. Likewise, it interprets problem resolutions into advice and explanations for the user. The knowledge base comprises facts and rules that pertain to the problem at hand. For example, a specialized knowledge base corresponding to an urban transportation network would contain facts regarding speed limits, adjacent land uses and land values, material and construction costs, grid layout and link capacities. In addition it might contain rules pertaining to grid design, physical relationships, engineering performance and policy interdicts. Other rules may pertain to tradeoffs between the cost of infrastructure improvements and increased congestion levels. Occasionally rules will conflict with each other and so a knowledge base also contains meta-rules for resolving those conflicts (Parsaye and Chignell, 1988).

The proposed system would draw on this knowledge base during its heuristic search for alternatives. Thus the knowledge base is essentially a database that contains the same facts and rules that an expert would use in determining how to restore service levels in the most cost effective way.

The other primary component of an expert system is the inference engine or control mechanism. This is a high level programming language based on boolean algebraic logical structures. Because expert systems combine this reasoning capability with a well structured body of facts and rules, they have become the most successful commercial application of artificial intelligence. A unique attribute of expert systems is that the control mechanism and the knowledge base are functionally distinct. This has encouraged the development of disembodied inference engines such as GURU (Micro Database Systems, Inc., 1987), known as expert system shells. These shells also provide a mechanism for introducing new facts and rules into the knowledge base.

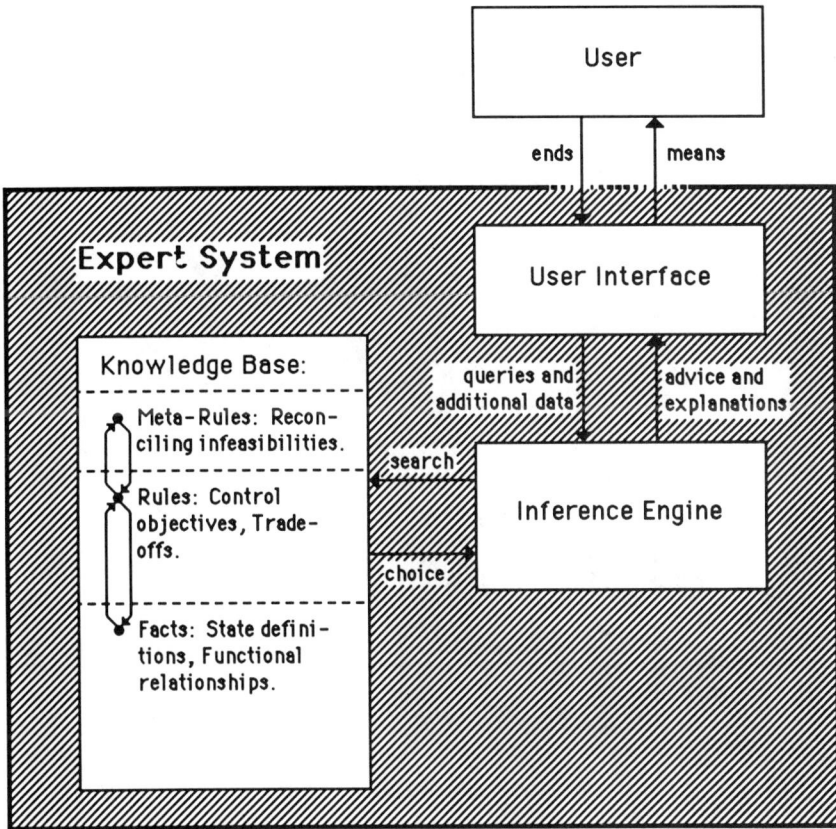

Figure 14.3. Components of an expert system (ES).

EGIS: Coupling ES and GIS

Figure 14.4 depicts the conceptual framework for an integrated expert geographic information system (EGIS). Here the ES judges which infrastructure improvements are appropriate within a particular context defined by the GIS production model. Put another way, ES can be used in conjunction with the GIS production model to organize a cost-of-mitigation model (Heikkila, 1988). It is this modeling approach that permits infrastructure improvements to be determined endogenously.

Consider the introduction of an exogenous change in demographic and land use conditions Δx. As before, the GIS production model (now resident

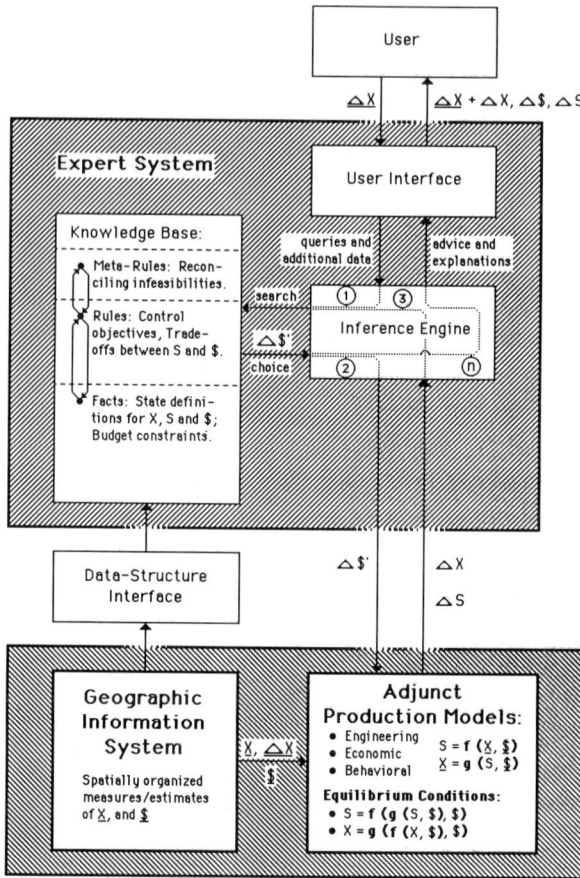

Figure 14.4. Components of an expert system geographic information system (EGIS).

within the EGIS) will produce spatially discriminated vectors of changes in transportation service levels Δs and in land use activities Δx that would result from these land use activity changes in the absence of infrastructure improvements. The inference engine of the expert system would then probe the knowledge base for facts and rules to identify a candidate change in infrastructure investments ($\Delta\$'$) for maintaining the target level of service provided by the system. The target level of service will also be determined endogenously through application of rules within the knowledge base governing the tradeoff between investment costs and service declines. The candidate investments ($\Delta\$'$) are spatially defined, as are their associated service impacts.

The adjunct production models then evaluate these proposed infrastructure changes within the context of the GIS. On the basis of this evaluation a revised vector of endogenous changes in land use activities (endogenous Δx) and service levels (endogenous Δs) is returned to the knowledge base for comparison by the inference engine with the operator's standing service level and cost objectives. If these objectives are not being met the heuristic decision rules that reside in the knowledge base are re-invoked to produce an alternative set of investments. These new investments would be identified in light of the fact that the previous alternative was unsatisfactory. Even if the target service levels were being met, additional searches may occur within well defined limits. In any event, the knowledge base (and possibly the boolean rules of the inference engine) would be updated with information on the performance of the inference rules employed, including whether or not a good feasible solution was found.

Incorporating EGIS into SCAG's Modeling Environment

As noted in section 2, the current SCAG modeling environment does not include EGIS. Under present conditions forecasting the general equilibrium location and transportation behaviors of households and firms would require extensive iteration between component models. Unfortunately, iteration implies expense, and even with a general equilibrium modeling capability, SCAG will not be in a position to undertake an exhaustive enumeration of the infrastructure alternatives it is routinely charged with evaluating. The alternative to an exhaustive enumeration is a systematically applied heuristic search, but at present expert judgement is applied to the forecast models in a rather informal or ad hoc fashion.

In a period characterized by local infrastructure shortfalls of every type, the social payoffs of effective infrastructure management are enormous. In the context of general equilibrium forecasting efforts, therefore, we submit that an expert system is most appropriate. Figure 14.5 illustrates how this can be accomplished within the SCAG modeling environment.

Consider, for example, the analysis of an anticipated population increase. The expert system feeds the projected population increase to the GIS which, in

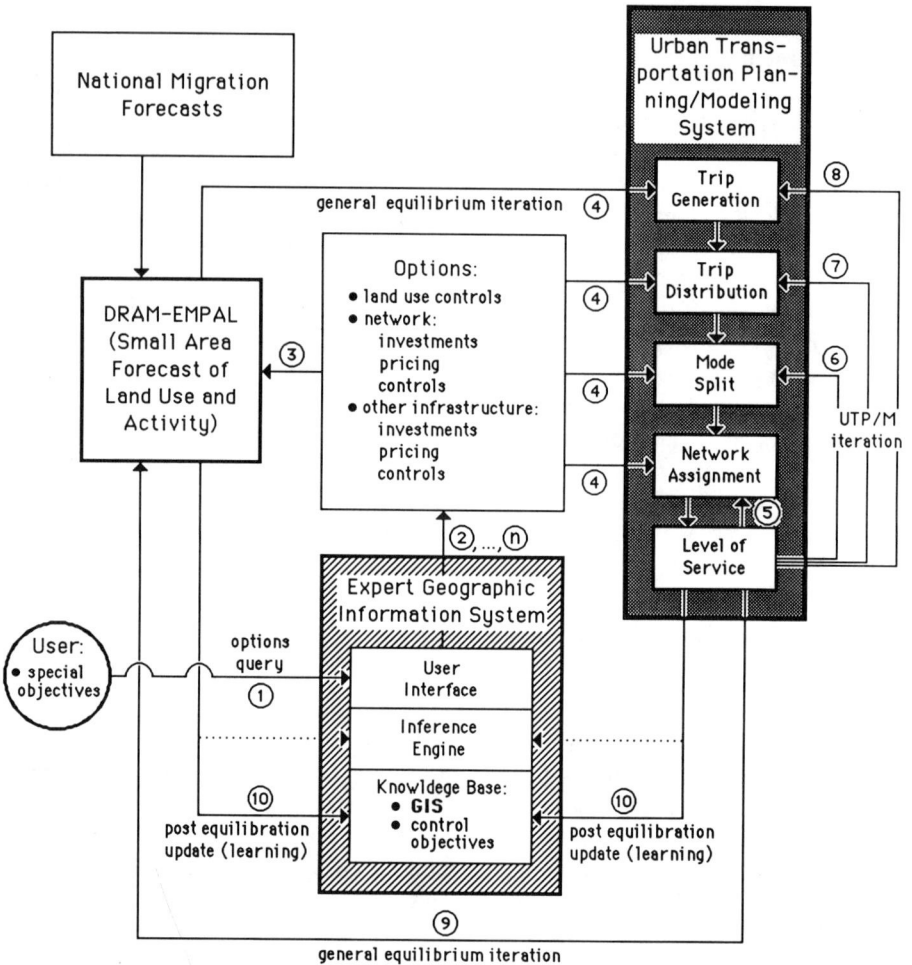

Figure 14.5. Using an expert geographic information system to formalize the search for infrastructure improvements.

conjunction with existing UTP models, determines that the existing transportation grid cannot provide the specified level of service. Drawing on a knowledge and decision base established for this purpose, the expert system executes an efficient exploration and evaluation of alternative transportation network investments. The outcome of this process is an estimate of the minimum expenditure required to maintain target transportation flows in light of the projected population increase. And as in Figure 14.4, the target service level is itself determined via rules governing the tradeoff between investment costs and congestion. That is, the output is an assessment of the mitigation and congestion costs associated with this population growth, based on a systematic application of the UTP and small area forecast models currently used by SCAG. These impact estimates are based on a general equilibrium conception of how land use activities, transportation service levels and infrastructure configurations will respond to exogenous shifts in demographic or land use variables. In this sense the EGIS model captures the essence of the general equilibrium conditions specified in Equations 5.

Notice that the generic EGIS model shown in Figure 14.4 adapts quite readily to the particular application at hand, that of SCAG. The next section discusses the extent to which this generic EGIS modeling approach may be even more generally applicable.

General Feasibility of EGIS for Local Government Planning Agencies

The preceding sections describe a framework for applying a generic method (EGIS) in a particular modeling environment (SCAG). This hypothetical EGIS has four attributes that must be described in detail before this approach can be adapted by local government planning agencies for infrastructure management purposes. These four areas pertain to (i) the land use and infrastructure models the EGIS uses to refine alternatives, (ii) the GIS data structure, (iii) the characteristics of the expert system knowledge base, and (iv) the input-output specifications of the system.

In several respects SCAG's modeling environment is more advanced than that of most local government planning agencies. Its GIS is well supported and is well integrated into the agency's ongoing planning and forecasting efforts. Moreover the partial equilibrium forecasting methods SCAG is striving for are exemplary. In sum, SCAG's current modeling environment combines with the Association's research agenda to address area (i) uniquely well. Notwithstanding these considerations, there is little reason to think that this modeling approach is beyond the ken of medium sized planning agencies focusing on appropriate problems.

Areas (ii) and (iii) are closely related because of the necessary complementarity between the GIS and the ES in an EGIS modeling approach. It is probable that any existing GIS database would require some modifications ancillary to the requirements of an expert system component. The expert system knowledge base presents a very significant challenge in any planning application. What is required are the same kinds of data that are used by public officials, municipal engineers and senior planning staff to evaluate alternative plans for expanding infrastructure capacity. This includes information on existing land uses, construction costs, existing supply and demand relationships and policy based interdicts bearing on expansion alternatives. Also, a set of decision rules for determining how these considerations should be applied in the search for a good feasible alternative must be obtained from the knowledge sources identified above and then encoded into the expert system.

These rules must ultimately be elicited from human experts, and the elicitation procedures involved must be basic yet consistent with the literature supporting existing decision analysis procedures for expert systems. Likely sources of experts include academics with specialized expertise in transportation systems modeling together with local government engineers and administrators. The rules defined for any specific problem context will partially dictate the content of the system's interface, though the form of this interface would be largely determined by the choice of expert system shell. In and of itself, a formal encoding of the assumptions and heuristic procedures already in use within the relevant planning agencies will promote improved decision making.

Communication between system components (iv) is a crucially important aspect of the EGIS design. Part of the expert system database must be self contained, notably the logical rules used for evaluating alternative decisions. The expert system will be responsible for guiding the GIS and supporting land use and engineering models in a coordinated manner, and the expert system must communicate with these other system components as it undertakes its deliberations. The expert system is also responsible for guiding interactions between the user and the system as a whole. Thus, its input-output specifications must be developed with careful intent.

On the basis of these considerations the potential for incorporating EGIS into local government transportation planning efforts may be stated as follows: probably feasible, but by no means trivial. A planning agency that might be considering such an undertaking would need to carefully evaluate its own circumstances to determine whether the expected benefits from more efficient infrastructure management outweigh the costs of incorporating EGIS. We are led to conclude that EGIS modeling of the impacts on infrastructure brought about by urban change is likely to be most suitable for planning agencies (i) that are responsible for making critical decisions regarding infrastructure improvements, (ii) whose planning jurisdictions are experiencing rapid growth or change and where those changes are leading to obvious strain on existing infrastructure capacity, (iii) whose current modeling environments are reasonably sophisticated and supported by a well trained staff, and (iv) who have the administrative and/or political will to undertake to encode their implicit or ad hoc judgements into a more formalized structure.

The potential benefits are enormous given the magnitude of infrastructure expenditures required in large urban settings: SCAG recently estimated that transportation mitigation measures for anticipated growth in Southern California over the next two decades would require expenditures of $42 billion. In this context, even small efficiency gains could obviously be of substantial value. The EGIS method outlined here appears to be feasible and worthwhile on the basis of a priori reasoning.

Acknowledgements. The authors wish to thank Thomas Baerwald, Edwin Blewett, Arthur Getis, Michael Goodchild and Hanlin Li for helpful comments on the earlier draft. Eric Heikkila gratefully acknowledges support from the Lincoln Institute of Land Policy and the USC Faculty Research Innovation Fund that enabled him to undertake this research. We are also indebted to Arnold Sherwood and his staff at SCAG for their splendid cooperation. The usual disclaimer applies.

References

Crosswell, Peter L., 1986. "Developments in data transfer between geographic information systems and mainframe computer databases," *Proceedings from the 1986 Annual conference of the Urban and Regional Information Systems Association*, Vol. II. pp. 47-61.

Duda, R., P.E. Hart, N. J. Nilsson, R. Reboh, J. Slocum, and G. Sutherland, 1977. "Development of a computer based consultant for mineral exploration," *SRI Report*, Stanford Research Institute, October.

Dueker, Kenneth J., 1987. "Geographic information systems and computer-aided mapping," *Journal of American Planners Association*, pp. 384-390.

Goodall, Alex, 1985. *The Guide to Expert Systems*, Oxford, England: Learned Information Ltd.

Han, Sang-Yun and T. John Kim, 1989. "Can Expert System Help with Planning", *Journal of the American Planning Association*, 55(3): 296-308.

Heikkila, Eric J., 1988. "Using ES and GIS to Evaluate Fiscal Impacts: Theory and Strategy", paper to be presented at the 1989 meetings of the Western Regional Science Association (currently under review).

Micro Database Systems, Inc., 1987. *GURU Reference Manual*, Lafayette, IN: Micro Datebase Systems.

Moore, James E., 1986. "Linearized Optimally Configured Urban System Models: A Dynamic Mills Heritage Model with Replaceable Capital," Ph.D. dissertation, Department of Civil Engineering, Stanford University.

Nelson, W.R., 1982. "REACTOR: An expert system for diagnosis and treatment of nuclear accidents," *AAAI Proceedings*.

NCGIA. Research Plan and Call for Participation, National Center for Geographic Information and Analysis, University of California at Santa Barbara.

Putman, Stephen H., 1977. "Calibrating a Disaggregated Residential Allocation Model - DRAM: The Residential Submodel of the Integrated Transportation and Land Use Package - ITLUP," *London Papers in Regional Science*, Vol. 7.

Raster, Peuquet D., 1978. "Data handing in geographic information systems," *Proceedings from the 1st Int'l Symposium on Topological Data structure for GIS*, Massachusetts: Harvard Lab for Computer Graphics and Spatial Analysis.

Silverman, B. G., 1987. "Should a manager hire an expert system?," *Expert Systems for Business*, Massachusetts: Addison- Wesley Publishing Company.

Suh, Sun-duck, Moonja Kim and T. John Kim, 1986. "An expert system for manufacturing site selection," *Planning Papers*, Department of Urban & Regional Planning, University of Illinois at Urbana-Champaign.

Waterman, Donald, 1986. *A Guide to Expert Systems*, Massachusetts: Addison-Wesley Publishing Company.

CHAPTER 15

Intelligent Urban Information Systems: Review and Prospects[1]

Sang-Yun Han and Tschangho John Kim

The salient characteristics of the computer, its enormous capacity and speed to store, access, and process data, has made it an indispensable tool to urban planners who deal with various information for their problem solving tasks. In urban planning, however, the role of the computer has been less significant than in other disciplines, such as business and engineering. As one possible explanation, Dueker (1982) cites that the distinctive nature of data used in urban planning—public goods and services—is that they are indivisible and therefore more difficult to describe discretely for computer processing.

Nonetheless, as artificial intelligence (AI) research shows the possibilities of developing an intelligent computer which is able to reason, learn, and understand human language, and as the planner's tasks become more complex, the role of computers and computerized information systems becomes increasingly important to the urban planning field.

The study of urban information systems has been primarily focused on devising an effective way of integrating data from various sources to provide the necessary information for decision making. Recent advances in artificial intelligence technology raise interest in how existing urban information systems might benefit from AI. Artificial intelligence is not a technology that can alone solve urban planning problems. Rather, it is a newly emerging and promising technology which can be incorporated or integrated into existing urban information systems to provide more intelligent and effective solutions to urban problems.

The objectives of this chapter are threefold: (1) to review the use of information systems in urban planning, (2) to examine the possibility of integrating

---

1 The original version of this chapter was published in *Journal of the American Planning Association*, 1989, 55(3):296-308.

AI technologies into existing urban information systems by reviewing the ideas of coupling expert systems and existing information systems, and (3) to examine some research issues in expert system application to urban planning.

The Role of Information in Urban Planning

The role of information in planning can be defined in various ways. For instance, Harris (1987) asserts that urban planning starts with data and information, because these sources describe the conditions of the real world from which planners try to achieve the aspirations and goals of society. In this context, the purpose of gathering, processing, and organizing data and of producing information is to understand the environment where the complex planning activities take place. Thus, better understanding of the environment through the use of information facilitates better planning.

Hopkins and Schaeffer (1985) treat planning itself as an information-producing activity that is performed to reduce the inherent uncertainties in decision making. Here, the major role of the planner is to produce and analyze quality information to aid effective decision-making on public policies. Catanese (1979) and Harris (1987) further clarify the meaning of information by constructing a hierarchy of data as illustrated in Figure 15.1. They emphasize a progression from observation to data, data to information, and information to intelligence. The diagram shows a vertical and circular feedback loop signifying that any level in the hierarchy may have problems which can be resolved at another level. In the diagram, knowledge represents the total concepts of data, information, and intelligence with a feedback loop.

Data may be explained as one-to-one relationships between observations and real world phenomena stored in a cleaned form. Information, on the other hand, is organized data which is a result of aggregation, manipulation, and other statistical, mathematical, or algorithmic changes of data. Information is derived from data to develop a certain knowledge necessary to solve a problem or to show patterns and directions.

As a level higher than information, Catanese (1979) explains intelligence as an ability to catch the essential factors from complex information and data. Harris (1987) defines intelligence as a product of interpreting data as a guide to actions. Intelligence, rather than data or information, may be more useful to planners in the decision-making process.

Comprehending the distinction between formal and informal information also aids understanding of the nature of information. As defined by Burch, Strater, and Grudnitski (1979), formal information is the information identified and formalized by society and its institutions regarding how, when, and for whom the information will be produced from data. Formal information in planning may include various legal requirements, planning procedures, and problem situations. Informal information, on the other hand, includes the information

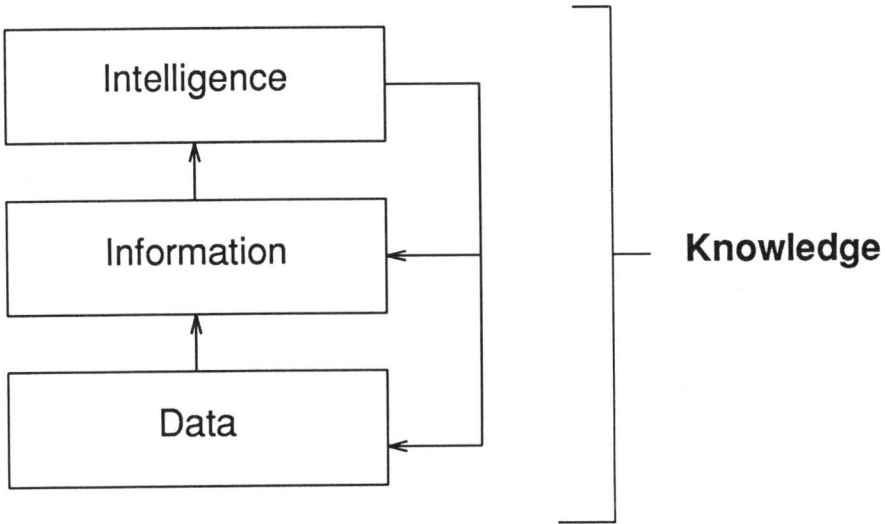

Figure 15.1. Hierarchy of Knowledge. Catenese (1979).

obtained through nonformalized procedures, such as personal judgments, hunches, intuition, hearsay, and personal experiences. The value of informal information is solely determined by the user of information.

Although informal information is needed to supplement formal information, the subjective and nonformalized nature of informal information has excluded its use in the domain of urban information systems. This is partly due to the technical limitations on conventional information systems. The use of informal information in urban information systems is discussed later in more detail. By examining the evolutionary process of urban information systems from simple data processing systems to intelligent urban information systems, the following section reviews how the transformation from data to intelligence can be achieved in urban planning.

The Nature of Urban Information Systems

In studying the use of information systems in urban planning, many definitional questions arise. In the planning field, information systems have gone by various names. Among them are "land information systems," "planning information systems," "environmental or natural resource information systems," and "geographic information systems." A generic term is needed that covers these different types of computer-based systems. For this purpose, "urban information system" might be a useful term. Most planners are familiar with this phrase, and it is a comprehensive term that covers all areas and functions of computer based systems built for planners.

In a generic form, urban information systems (UIS) may be defined as a formalized computerbased system capable of integrating data from various sources to provide the information necessary for effective decision making in urban planning. There is a class of systems known collectively as UIS. Classifying urban information systems based on the subject they deal with, such as "land information system" or "streams information system," is not as helpful in understanding the nature of UIS as classifying the UIS according to the functions and technologies involved in the system.

Based on that approach, four basic types of computer based information systems may be identified in planning. These include database management systems (DBMS), geographic information systems (GIS), decision support systems (DSS), and expert systems (ES). The major functional areas and characteristics of these systems are summarized in Table 15.1.

Table 15.1. Characteristics of the four types of UIS

UIS Type	Inputs	Processes	Outputs
DBMS	Raw data	Organizing, and modifying data and simple statistics	Processed data and customized reports
GIS	Point, line, and area data	Organizing and modifying data, geometric manipulation of data (cartographic modeling)	Composite overlay, graphic display of spatial data, customized reports
DSS	Raw and processed data, and models	Data analysis, operations research, and modeling other modeling	Information such as optimal values and other inputs to difficult decisions
ES	Facts and coded expert knowledge	Inferences and reasoning	Acceptable solutions and advice to judgmental problems

Database Management Systems (DBMS)

Traditionally DBMS has been primarily concerned with data storage, processing, and retrieval. In this chapter, DBMS is used interchangeably with traditional electronic data processing systems. In planning, the major purpose of
DBMS is to make data readily available to the planners in an orderly, efficient,
and effective manner.

In computer science, DBMS is defined in a very narrow sense as computer software that is used to manage data (Larson, 1982; Kroeber and Watson,
1987). In the planning discipline, DBMS may be defined in more general terms
as a computer based system which is capable of storing, updating, organizing,
and reporting data to the user in a timely, consistent, and efficient manner.
According to this definition, software such as dBase or Rbase are not DBMS
themselves, but rather fourth generation languages used to build DBMS.

One major function of DBMS in planning is to computerize routine tasks
of planners to enable fast and correct processing of data. Examples of these
tasks include organizing, updating, and reporting social indicator data or property transaction data. In addition to the recordkeeping function, most DBMS's
provide the capability of computing basic statistics.

In the business environment, DBMS often exists as a stand-alone system
to automate enormous transaction processing tasks such as daily sales and
inventory recordkeeping. In planning, however, DBMS tends to serve as a basis
for the other types of computer-based systems, including GIS and DSS.

Geographic Information Systems (GIS).

GIS may be thought of as a DBMS in that the basic function of both systems is
data processing. The type of data GIS deals with, however, is relatively unique;
and now advanced GIS provides more sophisticated data manipulation functions
for the various types of spatial analysis. Therefore, a more precise description
of the relationship between GIS and DBMS may be that DBMS is a part of GIS.
In fact, Raster (1978) argues that the core of any GIS is the database management systems.

Because of the capability of GIS to display spatial data in graphic form,
the term GIS is often mistakenly used as an interchangeable concept with
computer-aided mapping (CAM). As Croswell (1986) insists, GIS encompasses
both mapping and geographic analysis which is supported by a computer based
system that has graphic as well as analytical capability; but the terms GIS and
CAM may not be used interchangeably.

The efforts made by Dueker (1987) to distinguish GIS from CAM systems
while identifying links between them are very helpful in understanding the
nature of GIS. In his words, CAM is display-oriented, whereas GIS is analysis-
oriented. In other words, GIS is capable of performing spatial analysis, which
requires geometric manipulation of data to see the relationships among the
objects in the same or different layers. Meanwhile, CAM systems merely overplot separate layers of data types and lack the capability of relating data across

layers.

Spatial analysis is a very important function of GIS. Teng (1986) explains spatial analysis as cartographic modeling and defines it as "the process of manipulating single or multiple sets of digital map themes." Cartographic modeling may be explained as a computerized version of the earlier manual overlay techniques (Steinitz, Parker, and Jordan, 1976).

Presently, advanced geographic information systems provide much more complex spatial analyses, such as estimating runoff volume in specific areas, locating areas with scenic amenity, and searching for paths through three-dimensional space that satisfy certain conditions, i.e., minimizing distance or construction costs and avoiding major obstacles.

Decision Support Systems (DSS)

The DSS can be regarded as a distinctive type of urban information system because of its unique structures (see Figure 15.2) and the unique type of problems it deals with. It may be asserted that DSS is an enhanced version of DBMS upgraded by the addition of model base. In fact, the output of DBMS serves as an input of DSS.

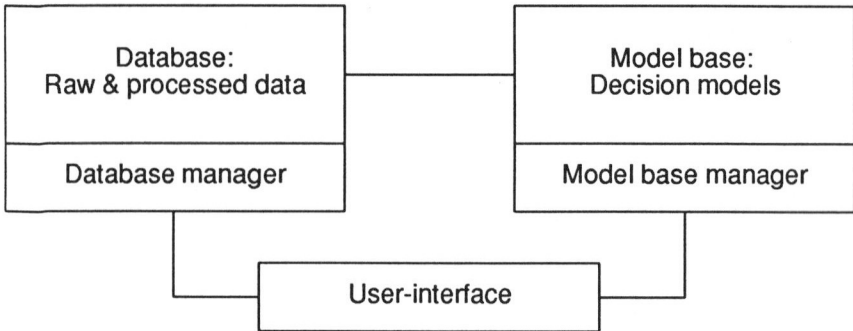

Figure 15.2. Basic components of a DSS.

The problems dealt with by DSS are generally different from those dealt with by DBMS. DBMS is suited for structured problems which have standard operational procedure, decision rules, and clear output formats, such as identifying low income districts or determining the median income of a city. DSS, on the other hand, is intended for unstructured or semistructured problems, such as evaluating land development proposals, for which DSS can be used to estimate fiscal and other impacts of proposal, providing quantitative support to the decision maker.

Kroeber and Watson (1987) define DSS as "an interactive system that provides the user with easy access to decision models and data in order to support semi-structured and unstructured decision-making tasks." As the definition implies, the interaction between the decision maker and the system is very important in DSS. The interaction is usually achieved in the form of "What-If" type dialogue. For example, when the user inputs different values of discount rate, the system produces new expected cost/benefit estimates, calculated using the cost/benefit estimating model in its model base.

As depicted in Figure 15.2, the three major components of DSS are (1) database, which contains raw and processed data, (2) model base, which contains various models including operations research (OR) models such as simulation or optimization models, and (3) user-interface, which enables easy interaction between the user and the system. As with GIS, the role of DBMS is important in building DSS, because DSS needs an input from DBMS to run its models.

The decision support system PLAN-DSS illustrates the use of DSS in planning to support decision making by systematizing the planning process. The purpose of PLAN-DSS is to analyze existing conditions of an urban area and forecast future trends. It maintains a database and database manager to handle data query and to supply necessary data to the model base. Its model base contains a collection of analytical models such as population projection, economic base analysis, and housing projection models (Djunaedi, Hinojosa, and Fowler, 1986).

With advances in the software development, the task of creating, storing, and accessing a model base can be easily achieved. Now many urban planning models can be built using several DSS generators, including spreadsheet programs and some statistical packages. Brail (1987) illustrates the ⌐ ￼of spreadsheet programs to create planning models, including a population allocation model, a Lowry model, an employment projection model, and an urban transportation model.

For unstructured and semistructured problems, planners can select appropriate models from the model base, run them using data in the database, and use the output from the DSS as a basis of their decision making. An example of this type of problem may include identifying sites for solid waste disposal. The decision making on this problem may be effectively supported by a DSS that maintains several models designed to estimate siting impacts such as environmental hazards and changes in the surrounding land value. This type of DSS can be more effective if supported by a database that provides necessary data to the model base.[1]

Expert Systems (ES)

The fourth type of UIS is the expert system, still regarded in urban planning as a new technology with many hopes. Expert systems have evolved as a branch of artificial intelligence and have been successfully introduced mostly in the field of medicine, chemistry, engineering, and the military.[2] In general, the expert system is defined as "a computer system that uses a representation of human expertise in a specialty domain in order to perform functions similar to those normally performed by a human expert in that domain" (Goodall, 1985). The essence of ES is that they attempt to incorporate the judgment, experience, rules-of-thumb, and intuition of human experts into problem solving. This knowledge of human experts was explained as informal information earlier in this chapter as knowledge is obtained through nonformalized procedures. The subjective and nonformalized nature of informal information has excluded the use of informal information from the domain of urban information systems, although it is needed to complement formal information.

The efforts made by several authors to distinguish expert systems from conventional programming are useful in understanding the nature of expert systems. According to Waterman, (1986), "an algorithmic method of conventional programming is designed to produce optimal solution whereas the heuristic method of expert systems produce an acceptable solution most of the time." Thus, the problems suitable for expert system development must require heuristic solutions rather than algorithmic solutions. Such problems may include finding new mineral deposits or settling a law suit in which heuristics and rules-of-thumb can effectively limit the search for solutions (Waterman, 1986).

Another distinctive characteristic of expert systems is their knowledge intensiveness. Newell (1985) points out that "an algorithmic program uses a small amount of knowledge (e.g., the knowledge of matrix multiplication) repeatedly over many cycles, whereas expert system typically has to search a large amount of knowledge at each cycle, and a particular piece of knowledge may apply only once." Another important feature of expert systems is the separation of the knowledge base or expert knowledge from the inference engine that applies the knowledge to the current problem. This feature enables planners to develop their own expert systems by taking advantage of expert system shells which provide a built-in inference engine. Another distinctive feature of expert systems is their explanation facility, which explains to the user the inference process used to reach conclusion.

As depicted in Figure 15.3, the generic components of expert systems include: (1) user interface, which allow the user to communicate with the system, providing necessary data to the system; (2) inference engine, which solves given problems using input data from the user and knowledge from the knowledge base, through its own reasoning methods; and (3) knowledge base, which contains the knowledge obtained from a domain expert, including facts, belief, rules of thumb, and other judgmental factors of the human expert. The inference engine and the facilities to develop knowledge base and user interface are usually provided with the commercially available expert system shells.

Regarding the possible application of expert systems in urban and regional planning, Ortolano and Perman (1987) provide a list of urban planning problems that can benefit from ES development. To date, only the developing ideas and concepts have been explored for the ES application in planning, with

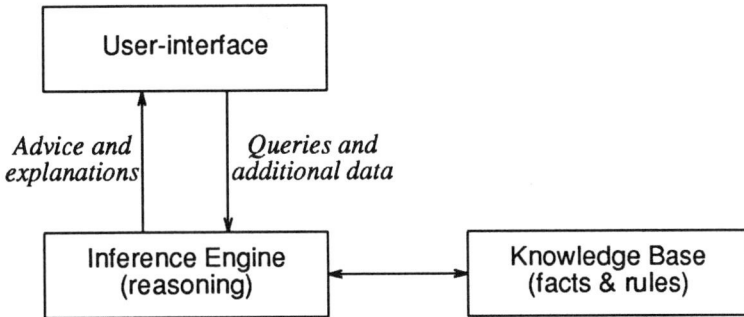

Figure 15.3. Basic components of expert systems.

some prototype systems being produced. Research on expert system application, in particular, has lagged in urban planning, while other disciplines have produced significant works with ES application.

There are several reasons for the slow pace of adaptation and development of expert systems in planning. One fundamental reason might be disparities between the type of problems planners deal with and the type of problems for which the problem-solving approach of expert system is suited.

For this issue, the limitations of the ES approach and the characteristics of the problems suitable for ES application need to be examined. Waterman (1986) provides useful guidelines for selecting suitable tasks for expert system development. Among them:

- Genuine experts exist and experts can articulate their (problem solving) methods.

- Experts agree on solutions.

- The task is not poorly understood.

Silverman (1987) provides additional guidelines:

- The problem typically takes a few minutes to a few hours to solve.

- No controversy over problem domain rules exists.

Further, Goodall (1985) suggests:

- The problem is clearly specifiable and well-bounded.

In addition, we suggest that:

- The problem solving should be judgmental in nature, not numerical.

Let's briefly examine these requirements. Are there genuine experts in the urban planning field? In expert systems, the term expert is normally used to represent a domain expert who has a high degree of knowledge in a very specific area. Because of the distinctive nature of urban planning, practicing professional planners must deal with multidisciplinary activities that embrace social, economic, political, and even anthropological factors. Also, problem-solving methods used in planning tend to be multidimensional, and are relatively difficult to articulate. Thus, the first requirement is not quite satisfied. Regarding the second guideline, it is relatively hard for planners to agree on a solution to a specific problem, because each planner may have his own philosophy and ideals. For the same reason, the fifth guideline is not well satisfied. As for the third, fourth, and sixth guidelines, it is very difficult to find a urban planning problem which is well-bound, well understood, and takes a few minutes to a few hours to solve.

The process of selecting a landfill site illustrates this problem well. The selection task is not a matter of selecting a site with minimum or acceptable environmental hazards. The selection process usually involves political processes that aim to resolve conflicting interests of many different groups. It can take a month or several years. And even if planners are entitled to select a site with minimum environmental hazards, there are too many variables which are not value-free.

As for the final guideline for expert system development, it is true that many planning problems require judgments of planners for problem solving. But our judgments are pretty subjective compared to the physician's judgment used to diagnose a disease. It is difficult to make value-neutral and rational decisions in urban planning solely based on our judgments.

On the other hand, other disciplines which have produced successful expert systems have many suitable tasks which satisfy those guidelines and requirements. For example, MYCIN, one of the most successful and pioneering expert systems, has only about 400 If-Then rules in its knowledge base. It diagnoses bacterial diseases with ninety-seven percent accuracy outperforming human physicians; and DENDRAL effectively finds the structure of organic compounds with 445 rules (Stepp, 1987). The task of diagnosing bacterial disease or finding the structure of organic compound is well bounded and performed by a genuine expert, competent physician or chemist in a few hours. If any controversy exists over the problem domain rules, it may be resolved by a few hours of discussion.

Facing these limitations on applying expert systems in urban planning, three hypotheses may be developed regarding the use of expert systems in urban planning. The first hypothesis is that the nature of urban planning problems makes expert systems intrinsically unsuitable in urban planning. Based on this hypothesis, the only thing planners can do is to spend much more time on identifying a few suitable urban planning problems which satisfy the basic requirements of ES development. One of the few urban planning problems which satisfies those requirements may include the task of interpreting land use laws and other legal issues to review land development proposals. In fact, in the law and business field, ES has been well applied to problems that involve rules of thumb, reasoning, and other non algorithmic approaches.

The second hypothesis on the use of ES in urban planning is that the limitations on the application of expert system in urban planning is mainly due to the technical as well as theoretical limitations of expert systems which can be solved with further researche. To elaborate on this point, the final section of this chapter is devoted to the current research issues in expert system development.

The third hypothesis is that most urban planning problems are composed of several subproblem areas, which have their own suitable technologies to be adopted. Based on this hypothesis, only the integration of different technologies can effectively solve urban planning problems. The idea of coupling expert systems with other types of information systems to create a more effective and intelligent system has recently garnered attention (Kowalik, 1985; Robinson et al., 1986; Kerschberg, 1987). This trend is due not only to the practical limitations of ES applications but also, more importantly, to some unique capabilities of ES that can benefit other systems. The notion of coupling different types of urban information systems is discussed in the following section. It is the purpose of this discussion to provide some insights into the development of more intelligent and effective urban information systems.

Developing Intelligent Urban Information Systems

To use the term, "intelligent urban information system," the word "intelligence" needs to be defined. While artificial intelligence is commonly defined as the study of ideas that enable computers to be intelligent, the word "intelligence" is used in many different ways. AI researchers define it as the ability to reason or the ability to perceive and manipulate things in the physical world (Winston, 1986). This definition may be useful in engineering fields such as robotics, but not in urban planning.

As discussed earlier, the definition of intelligence offered by Catanese (1979), who explains intelligence as one higher level of information, is more useful. He defines intelligence as an ability to catch the essential factors from complex information and data. In this regard, it may be asserted that expert systems exhibit some type of intelligence. They distill the essentials from complex information through judgmental capabilities similar to those of human experts.

Rather than trying to define "intelligence" more clearly, this chapter uses the term in a relative sense. That is, the system which communicates with the user in plain English, provides more understandable advice, explains the reasoning method used to obtain its solution, or possesses heuristic knowledge is more intelligent than the system which provides only query language interface or produces an optimum solution without showing the logic of the model.

The goal of developing an intelligent urban information system is to produce more useful computer tools that help solve urban planning problems not only by ordinary numerical computing but also by doing computing that exhibits intelligence. This goal may be achieved by effectively combining different types of technologies available, based on the notion that each system has its own areas of strength and one system can be supplemented by another. The following section discusses the ideas and concepts of coupling expert systems with several different types of information systems.

New Directions: Coupling Expert Systems with Existing Information Systems

1. Expert systems (ES) and database management systems (DBMS).

As evidenced at the First International Conference on Expert Database Systems (Kerschberg, 1987), the ideas of combining ES and DBMS have received a great deal of attention recently, creating a another hybrid of information system called "expert database system (EDBS)." The marriage of ES and DBMS takes two fundamental forms. The first type of coupling is exemplified by an 'expert' or 'intelligent' interface to a standard DBMS or an 'expert' query optimizer for standard DBMS. The second type of coupling includes a standard expert system integrated with a large relational database of facts (Brachman, 1987).

The functions of the 'intelligent' interface may include formulating efficient queries for the user by incorporating the knowledge of a human expert in the domain of the database, or interpreting data retrieved from the database and eliminating any inconsistencies, such as unifying the different units of measurement used in the database. The functions of the database system in the second type of coupling include the use of variables in the database system directly by the rules in the expert system as referred variables. In fact, the inability of an ES to access or utilize a database already developed has been one of the major deficiencies of ES.

Goodall (1985) points out some possible difficulties with the coupling of ES and DBMS:

1. an expert system and a database system are both big programs, making it difficult to run both together,

2. database systems may not be able to answer all the forms of query which an expert system would like to put to them, and

3. an expert system may not understand the replies by a database system.

But these problems have been solved recently by the introduction of several expert system shells, including GURU from Micro Database

Systems, Inc., which provides a database manager fully integrated into expert system environment (Micro Database Systems, Inc., 1987).

2. Expert Systems and Geographic Information Systems

As reviewed earlier, the major function of GIS is to store and manipulate spatial data for cartographic modeling. The basic need for coupling expert systems with GIS stems from the fact that many tasks involved in the cartographic modeling require expertise of the user on the particular subject matter, and that better expertise can be provided by experts or expert systems.

For example, the task of identifying suitable sites for a particular land-use can be effectively assisted by GIS, which easily identifies the area with desired or undesired characteristics through several overlay manipulations. For this task, however, the user must first define what is desirable and what is not. For instance, highly permeable soils may be undesirable for landfill sites, but desirable for other purposes. The selection criteria must be determined by the user prior to any GIS analysis and the criteria certainly can be more justifiable if developed by experts. Using the rule-based structure of expert systems, the knowledge of, for example, soil experts can be easily encoded into the knowledge base of expert systems.[3] Another potential contribution of expert systems to GIS is to help users of GIS to employ correct overlay combining methods that are logically valid. As explained by Hopkins (1977), the selection of proper overlay combining methods is critical in creating any valid final suitability map. It may be common for the novice users of GIS to apply invalid mathematical operations for overlay combinations, such as adding ordinal scale numbers.

The "rules of combination" methods explained by Hopkins (1977) can be a valid and effective method for combining overlays. These rules, however, must have "a logic based in the understanding of the natural system being described rather than a single set of relationships repeated for all combinations regardless of the specific types and factors being combined" (Hopkins, 1977). It implies that for non-experts it is difficult to develop a set of justifiable "rules of combination." In this regard, expert systems may be used to provide expert knowledge of "rules of combination" to GIS users. There is good potential that expert systems may be effectively linked to GIS to provide better input to GIS analysis when human experts are not available.[4]

Robinson, Frank, and Blaze (1986) identify an additional list of problem domains of GIS in which expert systems can be applied. These include automated map design, which emulates an expert cartographer in the task of placing feature names on a map using a heuristic graph-searching algorithm, and an automated feature extraction which detects valleys, streams, and ridges using the procedural knowledge encoded in the knowledge base. It should be noted, however, that the area of automatic feature extraction needs to be studied in close relation to AI research on computer vision. Recently several methods have been intensely studied in AI to understand visual processing. These include texture analysis, which finds surface patches that consist of similarly organized elements to extract information from the changes in these elements; stereo disparity analysis, in which the slight difference of using left and right eyes reveals the depth and orientation of nearby surfaces; and photometry analysis that derives the shape of surfaces from shading (Charniak and McDermott, 1986). The advances on computer vision and expert systems will certainly benefit planners who deal with remote sensing data, including aerial photos and Landsat data for landuse and land cover classification and monitoring.

In summary, the most common use of expert systems in GIS may be to provide useful and more intelligent interface to the user. Expert systems may be designed to help users devise efficient operating procedures for cartographic modeling and interpret the result of spatial analysis. Expert systems can also be used in GIS to enable uncertainty reasoning and to resolve inconsistent and contradictory information obtained from GIS. Among the problems of integrating ES with GIS, Robinson, Frank, and Blaze (1986) point out the limited capability of current ES shells and the lack of formalism typical in geography.

3.   Expert Systems and Decision Support Systems

The idea of integrating expert systems into decision support systems to create more powerful and useful computer-based systems also has been given much attention recently, creating a new terminology, "expert decision support system" (EDSS) or "intelligent decision support system." The possible contributions of ES in EDSS include: (1) helping users in selecting models, (2) providing judgmental elements in models, (3) simplifying building simulation models, (4) enabling friendlier interface, and (5) providing explanation capability (Turban and Watkins, 1986).

Most of all, the ES can play an important role in EDSS with model selection and building. As Strauch (1974) points out, the process of problem analysis (or modeling) usually involves three interrelated components: formulation of the formal problem, mathematical analysis, and interpretation of the results. While the mathematical analysis is well handled with DSS, the formulation requires the subjective knowledge of the user. Further, the interpretation requires the personal judgments of decision makers. The coupling of ES and DSS in this case is based on the assumption that subjective knowledge and personal judgment can be better if made by experts. An example of the use of ES for this purpose is the Advisory System for Ground Water Quality Assessment (Armstrong, 1987) which maintains a knowledge base in addition to the model base to assist decision makers in determining proper sample patterns and size given a specified confidence level and evaluating samples obtained.

Wood and Wright (1987) also show an example of adding a rule-based system in the simulation model for storm water management for the purpose of aiding users with model calibration and result interpretation.

The intelligent interface provided by expert systems for the modeling tasks in DSS may stimulate planners to employ mathematical models more frequently and easily in their problem solving processes. As often criticized by planners, modeling components tend to be treated as a black-box, inadequately recognizing the need for judgments by the users and concealing implicit judgments and assumptions from the users (Langendorf, 1985). The study by Wellman (1986) that attempts to connect expert systems to mathematical modeling is thus worth much attention. He reports a rule-based system that generates parameters for the user to make the mathematical models easy to use. In that system, ES serves as an extra layer between the model and the user, translating qualitative criteria into the numeric input, and also translating the model's numeric output to qualitative concepts that are more intuitive and informative to the user (Wellman, 1986). This approach certainly provides a great improvement over the unaided use of modeling algorithms, encouraging easier use of quantitative modeling to support many planning decisions.

In summary, the coupling of ES and DSS basically takes two different forms: (1) integration of ES into the conventional DSS to provide qualitative reasoning capability and intelligent user interface, and (2) integration of DSS into the conventional ES to provide modeling capability. In the first type of coupling, ES may help the users select proper models, input necessary parameters and interpret outputs of DSS. In the second type of coupling, DSS provides modeling capability to ES, recognizing that human experts often use quantitative models to support their experience, intuition, or rules-of-thumb. For example, site selection tasks require planners' intuition and experience to develop selection constraints and criteria, but the task of selecting the best site may require planners to use multiobjective optimization modeling to make more competent decisions.

Research Issues in Developing Expert Systems in Urban Planning

To examine the possible reasons for the less intensive application of expert systems in urban planning, this chapter earlier raised the issue of whether the limitations on the application of expert systems in urban planning is mainly due to the technical or theoretical limitations of the expert system itself, which may be solved by further research. To answer the part of this question, this section reviews major research issues in ES development.

Although the research on artificial intelligence began in the mid-1950s by a group of people who developed new concepts of symbolic processing, heuristic search, knowledge representation, and cognitive modeling, it is only in 1980 that expert systems have become a prominent subfield of AI research (Klahr and Waterman, 1986). Since expert systems are still relatively new, they create many research issues which need to be resolved for them to become practical computer tools. This section examines four major research issues, including uncertainty reasoning, knowledge acquisition, conflict resolution, and validation and evaluation of expert systems.

A. Reasoning with Uncertainty

Any expert system possesses a knowledge base, a repository of human expertise that, for the most part, is imprecise and uncertain. It is therefore essential that expert systems have the capability for representing and processing uncertain information in a correct way. Although the role of probability theory has been important in expert system development, the way in which expert systems handle uncertainty tends to be rather ad hoc. This results in severe criticism of the poor use of probability theory and other well-established techniques for uncertainty reasoning in expert systems (Hart, 1986). The apparent problems with applying the uncertainty reasoning method used in most expert systems to planning analysis are discussed by Johnston and Hopkins (1987).

The ad hoc approach to uncertainty reasoning in other disciplines, such as medicine and law, may be acceptable because their outcome domains are relatively well defined, with less likelihood of new situations. Any uncertainty reasoning methods that successfully replicate the behavior of human experts (doctors or lawyers) and that pass empirical tests are acceptable. In urban planning, however, the outcome domains of most problems are incompletely defined; and there is the likelihood of new situations (Johnston and Hopkins, 1987). To develop any valid expert systems in urban planning, we need to devise a consistent and justifiable logic that can effectively handle uncertainty problems in urban planning.

B. Knowledge Acquisition and Learning

Knowledge acquisition is often regarded as a major bottleneck in ES development. This is because the expert system can provide advice only as good as the knowledge it has in its knowledge base, while the knowledge acquisition process is difficult and very time consuming with currently available technology. Therefore, an effective way of acquiring knowledge needs to be devised to make the development of full scale expert systems feasible. The research on effective knowledge acquisition may be more urgent in urban planning, because (as discussed earlier) our domain of problems is relatively large, and difficult to confine to boundaries, and thus requires more knowledge to be encoded in the knowledge base.

One way of facilitating the knowledge acquisition process is to effectively link the expert system to an existing database system. The database can be regarded as a simple kind of knowledge base from which we derive beliefs and inferences. Further research is needed to devise an effective way of coupling expert systems and database systems for this purpose.

Another way of facilitating knowledge acquisition is to develop a knowledge acquisition aid that effectively captures the knowledge of an expert through interactive interviews, distills the knowledge, then automatically generates a knowledge base. The software called Auto-Intelligence supposedly helps experts capture their own knowledge by (1) identifying the structure of the knowledge, (2) discovering the relative importance of decision making criteria, and (3) classifying information and inducing knowledge (Intelligence Ware, Inc., 1987). This kind of tool can be helpful in building expert systems where human experts who supply knowledge to the ES perceive complex relationships or come to a conclusion without knowing how they did it and cannot articulate the methods they used. To effectively reveal the knowledge structure of human experts, Olson (1987) suggests using several cognitive methods, including some "indirect methods" that do not rely on the expert's ability to articulate the information that is used. The suggested methods need to be closely examined for their applicabilities in urban planning.

One of the most challenging ways of facilitating knowledge acquisition is to make computers learn by themselves. In fact, machine learning is a hot research area in AI with some progress made so far. There are several types of learning strategies studied in AI. Stepp and DeJong (1987) provide a taxonomy of computer learning strategies:

1. Rote learning - This is a traditional example of computer learning, that is, learning by being programmed.

2. Learning by being told - This is an another example of indirect learning enabled by knowledge engineers in most expert systems.

3. Learning by analogy - An example of this type of learning includes learning electric flow from hydraulic flow.

4. Learning from examples - Computers learn generalized class descriptions from sets of class examples that are arranged into categories by human experts.

5. Learning from observation and discovery - This includes finding patterns and regularities from a sample and finding underlying conceptual relationships.

The fifth type of computer learning has already been used for many scientific discovery problems, as shown by the system BACON (Langley, 1979), which is designed to discover polynomial relationships from raw numerical data. The possibility of "learning by example" by computers has also been demonstrated by the expert system PLANT (Michalski et al., 1985). As summarized by Stepp and DeJong (1987), the rule learning tasks of PLANT consist of several steps:

1. Plants experts list 35 features to measure in order to describe each diseased plant.

2. Data is collected for 350 sick plants thought to suffer from one of 15 diseases.

3. Plant experts divide 350 data vectors into classes, each having only one disease according to the experts' diagnosis.

4. An inductive learning program finds simple generalized rules to describe each of the 15 classes, and the 15 rules are put into the knowledge base of PLANT.

The diagnosis derived by the PLANT is reported to be 97 percent correct, compared to the 71 percent precision by human experts (Michalski et al., 1985).

Deriving rules from examples or observation by computers certainly is a very efficient knowledge acquisition strategy which enables the building of large scale expert systems by significantly shortening the time needed for knowledge acquisition.

C. Conflict Resolution

Most of the expert systems developed in urban planning have not explicitly provided the mechanisms that can handle the situation where conflicting knowledge, i.e., the same production rules with different objectives, can occur. In most expert systems, however, it is very typical that several conflicting or different rules apply to the same problem in the reasoning process.

Without conflict resolution strategies explicitly specified, the default strategy embedded in most expert systems will be applying the first rule encountered in the consultation process, as in ESSAS (see Chapter 9) and ESMAN (see Chapter 8). But this simple method is likely to ignore other rules that may produce a better solution, and also ignore potential benefits of using meta-knowledge.

Part of the reason that most expert systems developed in urban planning do not use meta-knowledge is provided by Goodall (1986) who asserts that the use of meta-rules inevitably make the knowledge base harder to read and understand because one meta-rule affects the sequence of other rules called and the effect is distributed throughout the rest of the knowledge base. Another important reason may be that meta-knowledge is simply difficult to obtain.

The expert system XCON (McDermott, 1982), which is designed to configure VAX computers, provides an example of simple conflict resolution mechanisms. The mechanism used in XCON is simply "use the most specific rules when conflicts arise." For example, if one rule states "IF there is a power supply specified, THEN ..." and another rule states "IF there is a power supply specified AND it is for 240 V, 50 Hz, THEN ..." and both rules are applicable to the same situation, the second rule is selected because it is more specific.

The simple conflict resolution method used in XCON does not look promising for the planning purposes. It is not reasonable enough to provide good solutions. A more advanced approach to this problem is to encode "meta-knowledge" in the knowledge base. Meta-knowledge is explained as "the knowledge about the use and control of domain knowledge in an expert system: (Waterman, 1986).

For an example of using meta-knowledge, consider the followings meta-rules which are the rule about rules describing how other rules should be used (Waterman et al., 1983):

— Use rules which employ cheap materials before rules which employ expensive one.

— Use rules contributed by an expert before rules contributed by a novice.

— Use rules that save lives before other rules.

One of the important benefits of using meta-rules is that "by utilizing meta-knowledge an expert system can choose among different inference methods that one which fits best a given situation, or it can exert control on the domain level by successively assigning goals to be pursued during system operation" (D'Angelo, 1985).

If well designed, the meta-rules may be useful in the land use planning context. For example, for the expert system which evaluates land development plans, the developers of the system can explicitly specify their primary concerns in dealing with conflicts and can change their primary concerns whenever they want to as the subject matters of the system change. For instance, when the expert system reasoning is related to the evaluation of a plan from a environmental perspective, a meta-rule which gives first priority to the protection of endangered species over any other knowledge can be useful. And when the reasoning is related to the evaluation of a plan from a legal perspective, a meta-rule which gives first priority to the rules which avoid complex litigation can be useful.

Although this meta-rule approach has the limitation that it requires the developers to explicitly express the order of priority through the whole knowledge base, this may be a very similar method to the way actual human experts or decision makers deal with conflicting knowledge. Nonetheless, the possible benefits of using meta-knowledge have not been rigorously researched in the urban planning field.

D.   Evaluation and Validation of the System

      Expert systems have intrinsic problems related to testing. Unlike many traditional computer systems, it is not easy to define the paths through a program in expert system making it difficult to test for completeness and correctness (Hart, 1986). And in most algorithmic programming, the validity questions may arise only regarding the assumptions on which models are based. Once the end user accepts the assumptions, it is not difficult to derive valid answers, i.e., correct optimal values. Expert systems, however, are designed to provide acceptable solutions using knowledge obtained from experts. Then it raises the question of who determines the level of acceptableness and how.

      This problem may not be so serious in some other fields. In the medical field, for example, the validity of an expert system which diagnoses a particular disease may be relatively easily determined by several doctors through field tests. In urban planning, however, it is difficult for planners to agree on the solution provided by expert systems, mainly because their subjective judgments can differ much from each other and the expertise used to develop the knowledge base tends to be extracted from various sources. For example, the expert system ESSAS, which is designed to determine the suitability of a particular site for the construction of military facilities, has extracted expertise from various sources, including a master planner in a military installation, numerous field manuals (FM), and safety and environmental regulations (see Chapter 9). Clearly it is difficult to determine the validity of this system because human experts who possess all the knowledge used in this expert system are rarely available.

      The distinction between the terms evaluation and validation may be helpful in understanding the necessary tasks required for the effective testing of expert systems. In Liebowitz's term (1986), evaluation measures the system's accuracy and usefulness, while validation determines whether the correct problem was solved. Thus the evaluation process involves the users who determine the utility of the system and the domain experts who determine the accuracy of any advice and conclusion.

      In some cases, the evaluation or validation methods developed in other fields may not be applicable to the expert systems in urban planning, because the subject areas covered by the expert systems in planning can be very different in nature, and the planning decisions made using expert system can affect a relatively broad range of people. The several evaluation techniques developed in other fields as described by Liebowitz (1986) must be examined for their applicabilities in planning. If not applicable, we need to devise our own evaluation/validation methods to produce any acceptable expert systems in planning.

Summary

As noted by Langendorf (1985), it is ironic that most computer models used by planners have been developed for structured problems, while most decision making in urban planning addresses semistructured and unstructured problems. The reason for this irony may be that the development of computerized information systems for the semistructured and unstructured problems has been technically difficult and expensive. Active research on computer applications in planning and the advances in computer technology will certainly allow planners to employ more computer based systems for their less structured problem-solving tasks.

This chapter examined how various types of information systems can help planners with effective decision making. To facilitate discussion, it categorized urban information systems (UIS) into four types: DBMS, GIS, DSS, and ES. The functional areas of these systems, however, can be overlapped. For example, many DBMSs can be classified as data-oriented DSS and many GIS's can be categorized as spatial-oriented DSS.

For the purpose of providing planners with insights into the development of more intelligent information systems, the second part of this chapter was focused on the research issues in the application of expert systems in urban planning and the ideas of coupling expert systems with existing urban information systems. It is believed that building an effective urban information system requires an understanding of the limitations of each type of computer based system. Further combining different types of information system technology can help to reduce the limitations of each system.

Notes

1. The effective integration of database into model base has become the most important research topic of DSS research. Liang (1985) examines several issues in this area.

2. Examples include PROSPECTOR (Duda et al., 1978) which aids geologists in their search for ore deposit, DENDRAL, which is developed to infer a compound's molecular structure from mass spectral and nuclear response data (Waterman, 1986), and REACTOR (Nelson, 1982), which diagnoses nuclear reactor accidents.

3. The knowledge of soil experts may be encoded using relational data structure. But it can be simpler and more efficient to encode it using a set of rules. For an example of such rules, IF soil-type = drummer OR dana OR ... THEN erodibility = high AND permeability = low AND ....

4. The system GEODEX (Chandra et al., 1986) illustrates the effort of coupling ES and GIS for the task of site suitability analysis. It enables the use of expert landuse planner's knowledge for site constraint development, while using a GIS package for geographic data processing.

References

Armstrong, Marc P., 1987. "A rule-based advisory system for ground water quality assessment at hazardous waste disposal sites," *Working Paper*, Department of Geography and Computer Science, The University of Iowa.

Brachman, Ronald J., 1987. Tales from the far side of KRYPTON: Lessons for expert database systems from knowledge representation. *Proceedings From the First International Conference on Expert Database Systems*, Kerschberg, Larry. Ed. Menlo Park, California: The Benjamin/Cummings Publishing Company, Inc.

Brail, Richard K., 1987. *Microcomputers in Urban Planning and Management*, New Jersey: Center for Urban Policy Research.

Burch, John G., Felix R. Strater, and Gary Grudnitski, 1979. *Information Systems: Theory and Practice*, 2nd Ed. New York: John Wiley & Sons.

Catanese, Anthony J., 1979. "Information for planning," *The Practice of Local Government Planning*, Washington D.C.: International City Management Association.

Chandra, N. and W. Goran, 1986. "Steps toward a knowledge based geographic data analysis system," *Geographic Information Systems in Government*, edited by B.K. Opitz. Hampton, VA:

Deepak, A., Eugene Charniak, and Drew McDermott, 1986. *Introduction to Artificial Intelligence*, Reading, Massachusetts: Addison-Wesley Publishing Company.

Clarke, Keith C., 1986. "Advances in geographic information systems," *Computers, Environment, and Urban Systems*, 10: 175-184.

Croswell, Peter L., 1986. "Developments in data transfer between geographic information systems and mainframe computer databases," *Proceedings from the 1986 Annual Conference of the Urban and Regional Information Systems Association*, II: 47-61.

D'Angelo, Antonio, Giovanni Guida, Maurizio Pighin and Carlo Tasso, 1985. "A mechanism for representing and using meta-knowledge in rule-based systems," in *Approximate Reasoning in Expert Systems*, edited by M.M. Gupta et al. New York: North-Holland, 781-798.

Djunaedi, Achmad, Jesus H. Hinojosa and George C. Fowler, 1986. "PLAN-DSS: Using a microcomputer-based decision support system in an urban planning office," *Proceedings from the 1986 Annual Conference of the Urban and Regional Information Systems Association*, III: 57-69.

Duda, R., P.E. Hart, N.J. Nilsson, R. Reboh, J. Slocum, and G. Sutherland, 1977. "Development of a computer based consultant for mineral exploration," *SRI Report*, Stanford Research Institute, October.

Dueker, Kenneth J., 1987. "Geographic information systems and computer-aided mapping," *Journal of the American Planning Association*, Summer: 384-390.

Dueker, K.J., 1979. "Land resource information systems: a review of 15 years of experience. Geo-processing," 1(2): 105-128. Quoted by Keith C Clarke. 1976. "Advances in geographic information systems," *Computers, Environment, and Urban Systems*, 10: 175-184.

Dueker, Kenneth J., 1982. "Urban Planning Uses of Computing," *Computers, Environment, and Urban Systems*, 7: 59-64.

Fenves, Steven J., 1986. "What is An Expert System," *Expert Systems in Civil Engineering*, edited by Celal N. Kostem. New York: American Society of Civil Engineers.

Goodall, Alex, 1985. *The Guide to Expert Systems*, Oxford, England: Learned Information Ltd.

Harris, Britton, 1987. "Information is Not Enough," *URISA News*, 90.

Hart, Anna, 1986. *Knowledge Acquisition for Expert Systems,* New York: McGraw-Hill Book Company.

Hopkins, Lewis D., 1977. "Methods for generating land suitability maps: A comparative evaluation," *Journal of the American Planners Association,* October: 386-400.

Hopkins, Lewis D., 1985. "The Logic of Planning Behavior," *Planning Papers,* Department of Urban & Regional Planning, University of Illinois at Urbana-Champaign.

Intelligence Ware, Inc. 1987. Knowledge acquisition goes automatic. Advertisement on Expert Systems.

Johnston, Douglas M, and Lewis D. Hopkins, 1987. "Expert Systems in Planning Analysis: The Logic of Uncertainty," *Town Planning Review,* October.

Kerschberg, Larry. Ed. 1987. *Proceedings From the First International Conference on Expert Database Systems,* Menlo Park, California: The Benjamin/Cummings Publishing Company.

Kowalik, Janusz S. Ed., 1986. *Coupling Symbolic and Numerical Computing in Expert Systems,* New York: Elsevier Science Publishers B.V.

Kroeber, Donald W. and Hugh J. Watson. 1987. *Computer-Based Information Systems,* New York: Macmillan Publishing Company.

Langendorf, Richard, 1985. "Computer and Decision Making," *Journal of the American Planning Association,* Autumn: 422-433.

Leibowitz, Jay, 1986. "Useful approach for evaluating expert systems," *Expert Systems,* 3(2).

Liang, Ting-peng, 1987. "Expert Systems as Decision Aids: Issues and Strategies," *Working Paper.* Department of Accountancy, University of Illinois at Urbana-Champaign.

Liang, 1985. "Integrating model management and data management in decision support systems," *Decision Support Systems.*

McDermott, J., 1982. R1: "A rule-based configurer of computer systems," *Artificial Intelligence,* 1(19).

Michalski, R.S., J.H Davis, V.S. Bisht, and J.B. Sinclair, 1985. "PLANT: an expert consulting system for the diagnosis of Soybean diseases," *Progress in Artificial Intelligence,* edited by L. Steels and J.A.

Campbell Ed. New York: Ellis Horwood Limited. Micro Database Systems Inc., 1987. *GURU Reference Manual,* P.O. Box 248, Lafayette, Indiana 47902.

Nelson, W.R., 1982. REACTOR: "An expert system for diagnosis and treatment of nuclear accidents," *AAAI Proceedings.*

Newell, A., 1986. Private communication made by Fenves, Steven J. "What is An Expert System," *Expert Systems in Civil Engineering,* edited by Celal N. Kostem. New York: American Society of Civil Engineers.

Olson, Judith R., 1987. "Extracting expertise from experts: Methods for knowledge acquisition," *Expert Systems,* 4(3).

Ortolano, Leonard, and Catherine D. Perman, 1987. "A Planners Introduction to Expert Systems," *Journal of the American Planners Association,* Winter: 98-103.

Raster, Peuquet D., 1978. "Data handing in geographic information systems," *Proceedings from the 1st Int'l Symposium on Topological Data Structure for GIS,* Massachusetts: Harvard Lab for Computer Graphics and Spatial Analysis.

Robinson, Vincent B., Andrew U. Frank, and Matthew A. Blaze, 1986. "Expert systems applied to problems in geographic information systems: introduction, review, and prospects," *Computers, Environment, and Urban Systems,* 11(9): 161-173.

Silverman, B.G., 1987. "Should a manager hire an expert system," *Expert Systems for Business,* Massachusetts: Addison-Wesley Publishing Company.

Simon, H.A., 1960. *The new science of management decision,* New York: Harper and Row.

Steinitz, Carl, Paul Parker, and Lawrie Jordan, 1976. "Hand-drawn overlays: Their history and perspective uses," *Landscape Architecture,* September.

Stepp, Robert E. and Gerald Dejong, 1987. Classnotes from the course, Introduction to Artificial Intelligence. University of Illinois at Urbana-Champaign.

Srtauch, R.E., 1974. "A critical assessment of quantitative methodology as a political analysis tool, Santa Monica: Land," Quoted by Richard Langendorf, 1985. "Computers and Decision Making," *Journal of the American Planning Association,* Autumn.

Teng, Apollo T., 1986. "Toward a topologically integrated geographic information system for cartographic modeling," in *Proceedings of the Annual Conference of the Urban and Regional Information Systems Association,* Washington, D.C.: URISA.

Turban, Efraim and Paul R. Watkins, 1986. "Integrating Expert Systems and Decision Support Systems," *MIS Quarterly,* June.

Waterman, Donald, F. Hayes Roth, and D.B. Lenat, 1983. *Building Expert Systems,* Massachusetts: Addison-Wesley Publishing Company.

Waterman, Donald, 1986. *A Guide to Expert Systems,* Massachusetts: Addison-Wesley Publishing Company.

Wellman, Michael P., 1986. "Reasoning About Assumptions Underlying Mathematical Models," *Coupling Symbolic and Numerical Computing in Expert Systems,* edited by J.S. Kowalik. New York: Elsevier Science Publishers B.V.

Winston, Patrick H., 1986. *Artificial Intelligence,* 2d Ed. Reading, Massachusetts: Addison-Wesley Publishing Company.

Wood, David M. and Jeff R. Wright, 1987. "Calibrating Complex Simulation Models Using Rules-Based Inferencing," *Working Paper,* School of Civil Engineering, Purdue University.

Index